INSTA360®: HOW TO USE the INSTA360 X3

PUBLISHED BY **Kaisanti Press**

INSTA360: HOW TO USE THE INSTA360 X3.

FIRST EDITION: January 2023

Copyright

Trademarks

This book is in no way affiliated, authorized, sponsored, or endorsed by or any other companies mentioned in this book. All references to Insta360 and other trademarked properties are used in accordance with the Nominative Fair Use Doctrine and are not meant to imply that this book is an Insta360 product for advertising or other commercial purposes.

All terms mentioned in this book that are known to be trademarks or service marks have been appropriately capitalized. The author cannot attest to the accuracy of this information. Use of a term in the book should not be regarded as affecting the validity of any trademark or service mark.

Insta360® is a trademark or registered trademark of Insta360 and/or Arashi Vision, LLC in the United States and other countries. Any other trademarks remain property of the trademark holder.

Warning and Disclaimer

This book nor its author does not endorse any of the dangerous sports/activities shown in this book. If you choose to participate in these activities, do so at your own risk. This book does not guarantee that by following the advice in this book, your camera will not be damaged. Always refer to your camera's user manual. Always check with your camera manufacturer and follow recommendations before getting your camera wet. The author accepts no liability for any harm done to your camera or selves.

INSTA360®:

HOW TO USE the INSTA360 X3

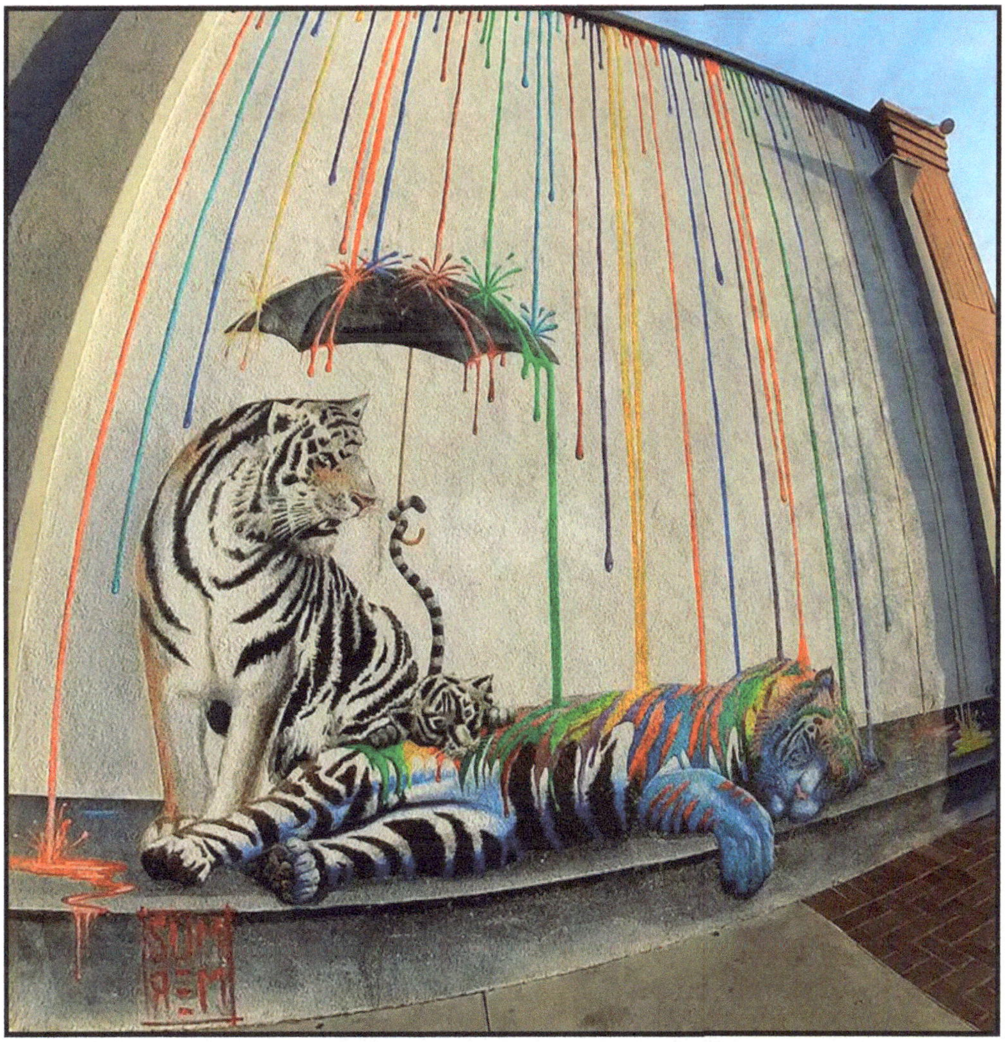

The Book for Your Camera- The Insta360 X3

by Jordan Hetrick

“The right perspective makes the impossible possible.”

INTRODUCTION

I first started teaching people to use action cameras more than a decade ago, when GoPro cameras first evolved into a true storytelling tool. I saw the potential of these revolutionary little cameras and knew people would need help to tap into the powerful features to really get the most out of their cameras.

I had already been using big, bulky water housings to take photos of surfers, bodyboarders and windsurfers around the world for years. The resulting images were worth the hard work, but when I got my first GoPro camera, the struggle of the big camera was gone and just the fun remained. That's when it all clicked. Everyone was going to want one of these, and people were going to need help. So I set out on a mission to figure out the clearest, most logical way for people to learn how to use their action cameras from start to finish.

When I began this journey, I didn't know the future, but I followed my heart. My help has been welcomed with open arms by people from all around the world.

Twelve years, thirteen books, and millions of readers/viewers later, you are reading the evolution of my intent to help you, written specifically for the Insta360 X3. Technology continues to evolve at a rapid rate and now you have the latest and greatest 360 camera available- the Insta360 X3.

This how-to-use guide will teach you how to use your Insta360 X3 camera with confidence from the initial setup all the way through to sharing your edited photos and videos. The technology has jumped leaps and bounds and it's all available right now… for you! Let me teach you everything you need to know. I'm so happy to have you on board!

Jordan Hetrick

Bestselling Author on GoPro Cameras

CONTENTS

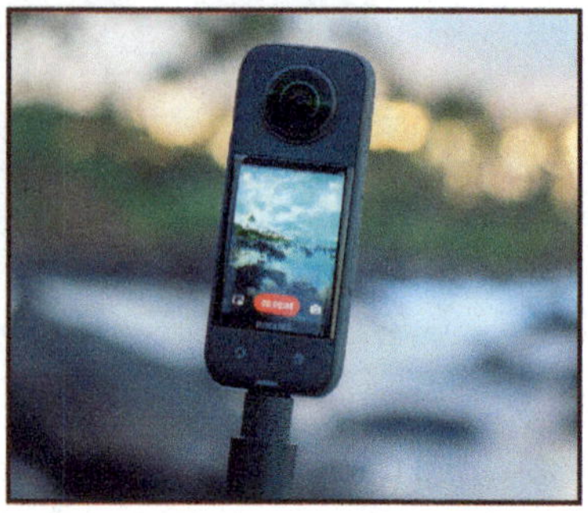

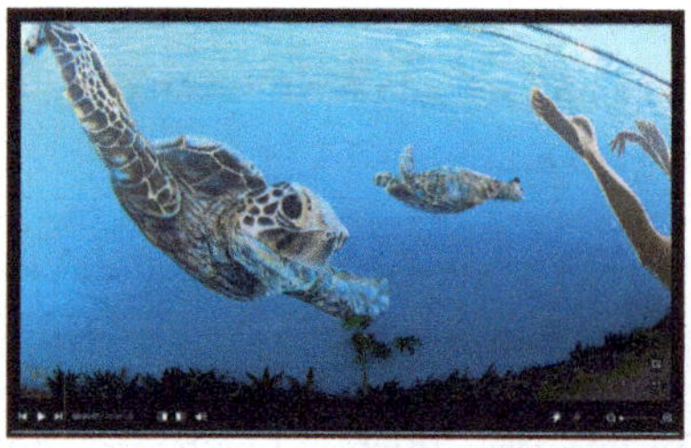

CONTENTS

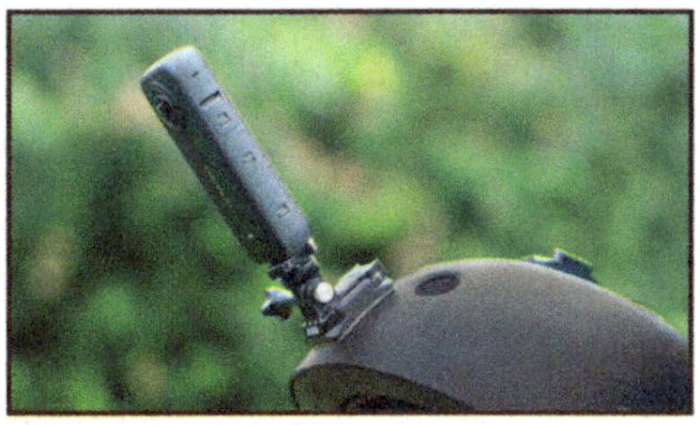

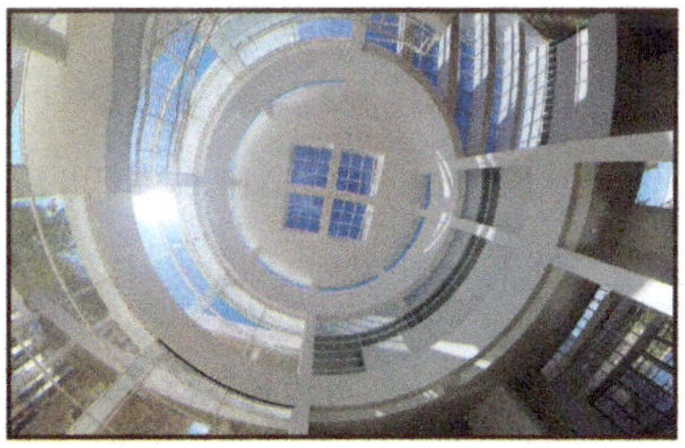

CONTENTS

ABOUT THE INSTA360 X3

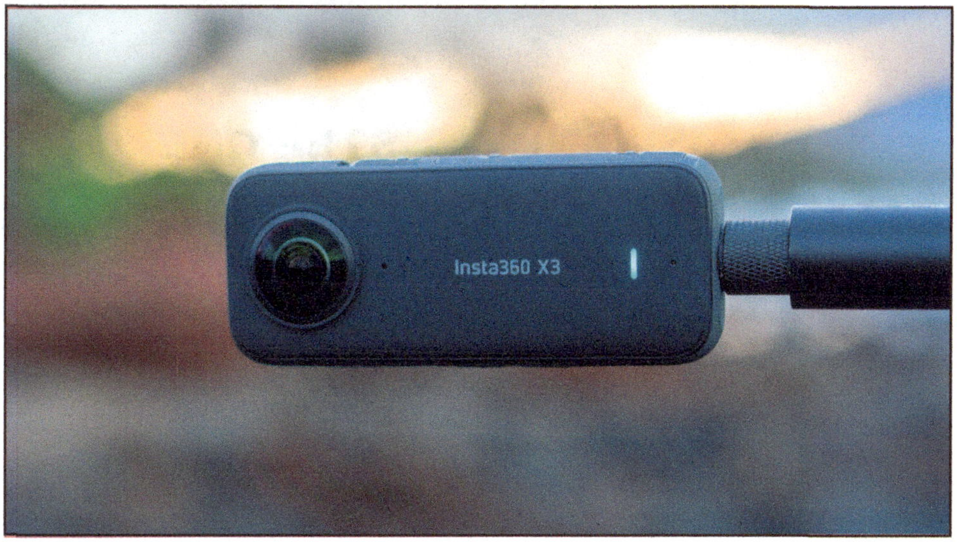

Welcome to the world of 360!
The Insta360 X3 is the next level of 360 cameras offering the ability to film life's magic moments in full spherical 360. Through creative engineering, Insta360 has created a camera that gives you so many possibilities for videos and photos like you've never seen before.

The Insta360 X3 uses two ultra wide-angle lenses to record everything into a full 5.7k spherical video or 72MP photo. This creates high quality 360 videos and photos, but, even if you don't like the look of spherical 360 photos or videos, the benefits of what you can create are mind-blowing. The Insta360 X3 is **a full-on one-person filming, editing and creating powerhouse.** Whether you are using the X3 for real estate, action sports, vlogging, or just to record life's simple moments, Insta360's X3 is THE tool that will create memorable content.

Let's talk about some of the magic you can create using the Insta360 X3. The Insta360 X3 gives you the technological power to: Create an invisible mount which makes the camera appear to be floating around you. Edit your photos and videos to a super wide-angle perspective for immersive action. Record full spherical 360 audio recording using 4 microphones. Film videos with unshakeable stabilization and horizon-leveling for footage you won't believe. Maximize in-camera stitching for an easier and quicker workflow. And create amazing transitions using your new knowledge.

You will learn about these techniques and way, way more in this book written for the Insta360 X3. This edition offers a fully comprehensive editing section, which will teach you how to use all of the tools you need to edit everything from a quick clip to a professional-looking video like a pro.

Being so unique, **there is a lot to learn and understand before you begin to love your Insta360 X3.** But with the possibilities of what you can create with the Insta360 X3, it's worth taking the time to get to know this camera.

There's never been a better time to record life's adventures, whether it's in your backyard, under the sea or on top of a mountain.

So grab your Insta360 X3 and let's learn everything you need to know about your new camera so you can create the best videos and photos ever! Once you get to know it, you are going to love it!

HOW TO USE THIS GUIDE

Now that you've decided to really learn how to use your Insta360® X3 camera, the information in this guide will teach you everything you need to know to get the shots you've always wanted.

This guide is organized into 7 Steps, which were written to logically **guide you through the learning process**. By the end of this book, you will have a clear and thorough knowledge of how to do everything possible with your Insta360 X3 camera.

In Step One- Get To Know Your Insta360 X3, you will learn how to **unlock the full potential of your new camera**. This section includes essential information to familiarize you with your camera and to get you started, as well as get you connected to the Insta360 App. You can also view the Insta360 X3 Online Manual on Insta360's website (search "X3 Basic Operation" in the search bar) on their website to be used in conjunction with the information in this guide. But as far as using the camera, this guide provides you with the vital knowledge to understand what you really need to know to use your Insta360 X3.

STEP ONE

GET TO KNOW YOUR INSTA360 X3

Learn To Navigate Your New Camera

Welcome to the Insta360 X3! This first section gives you hands on practice to easily navigate your new camera and access its features so you can focus on filming. Once you know your way around, you will be able to unlock the full potential of the Insta360 X3 to capture your exciting moments in the best way possible.

Let's get to know your new camera!

SETTING UP YOUR CAMERA FOR THE FIRST TIME (It's Easy!)

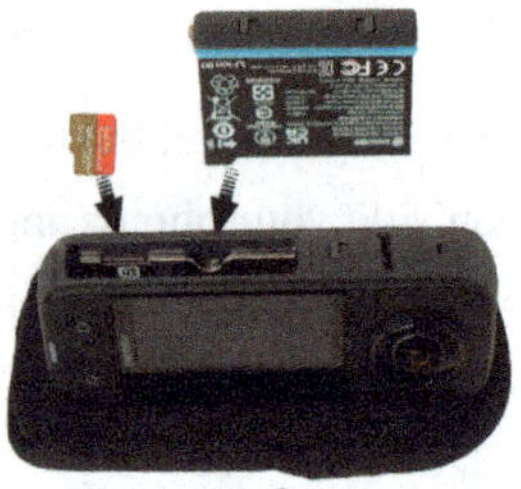

1-3

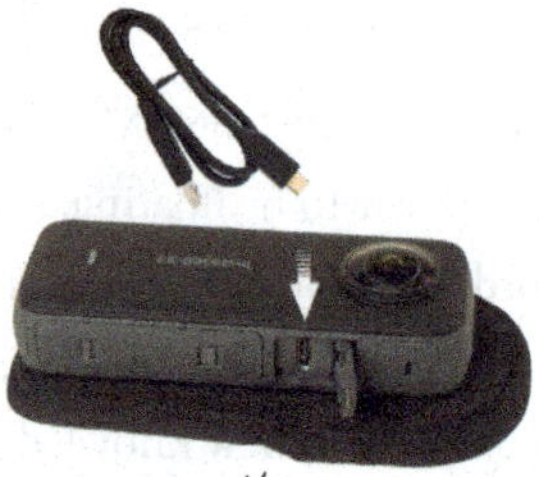

4

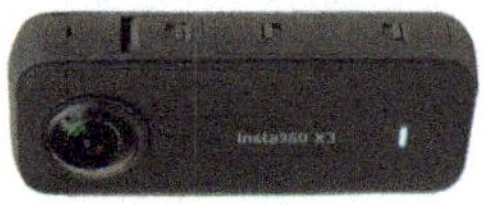

5-6

1. As soon as you open the box, set the camera on the carrying case to **protect the two lenses**. These can scratch and should not be set directly on any hard or rough surfaces.

2. Remove the battery and **insert the microSD card** into the slot. A 64GB microSD card is included in some X3 kits as a bonus. The X3 requires a **Class V30** rating / UHS-1 microSD card up to 1TB (such as the SanDisk Extreme Pro V30 A1 or A2). Use this exact class for compatibility.

3. **Reinsert the battery** with the 'UP' Arrow facing towards the lens. Make sure the battery is fully inserted with both latches closed without any orange showing. This has to be completely closed to stay waterproof.

4. **Charge your camera**. Open the side door above the battery and plug one end of the USB cable into the USB port on your camera and the other end to a computer or 5V/3A USB power supply. Your camera battery comes partially charged and using it with a partial charge will not affect the battery life. Charging typically takes about 2 hours using a wall charger or power pack and up to 4 hours using a computer.

5. Once your camera has charged, the light will turn off indicating that charging in complete. Remove the USB cable and **close the side door**, making sure the door closes completely. Slide the tab up so no orange is visible.

6. **Activate the camera.** Press the Power Button on the side of the camera to power on the camera. Follow the Setup prompts and connect to the Insta360 App. The camera has to be activated through the app before use, and the firmware should be updated to the current version.

Your camera is ready to use. It's that easy! Now let's go through elements of your camera.

Note: Keep the camera updated to the most current firmware to take advantage of the modes and settings as shown in this book. If your camera's firmware needs to be updated, the easiest way to update is to connect to the Insta360 App.

WHAT ARE THESE BUTTONS AND PORTS FOR?

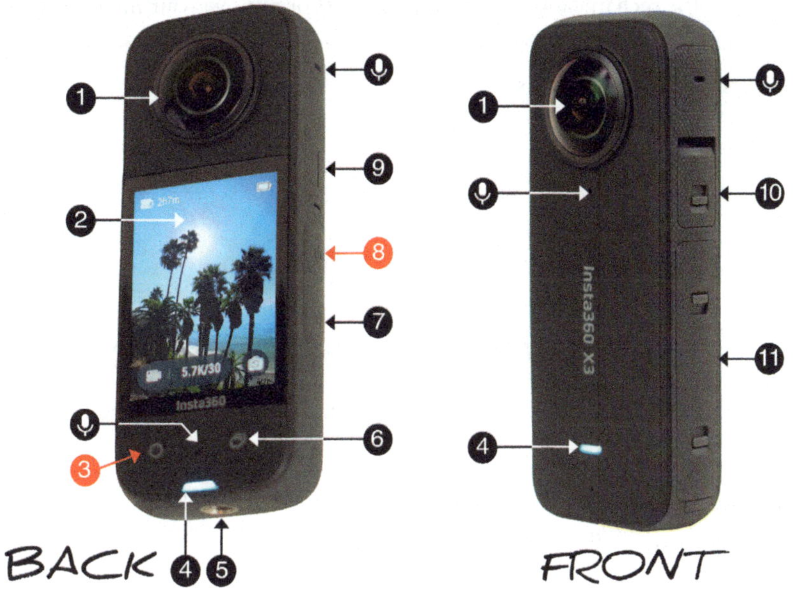

1. Lenses

The X3 has wide angle two lenses, one on each side of the camera, which are used for recording the full 360 degree view. These glass lenses are not replaceable, so it is extra important to keep these lenses free from scratches and dirt.

2. Touch Screen

The Touch Screen is a built-in LCD Screen that can be used as a viewfinder for composing shots, to change settings and to play back your footage. The Touch Screen section in this step gives you an explanation of everything you can do using this screen.

3. SHUTTER Button

Press this button to start and stop recording or to take photos. The camera beeps once when it starts recording. After recording a video, the camera beeps again three times after the file has been processed.

4. Camera Status Lights

These red lights turn on when the camera is recording video and when taking photos. You can turn off just the front light or both of the lights (one is also on the back) in the Preferences Menu.

5. ¼" Thumb Screw for Mounting

The ¼"-20 Thumb Screw is used to attach your camera to mounts. You will learn more about using setting up the X3 for creative mounting setups in Step 3.

6. Lens Button

Press this button to switch between the front lens, back lens, or both lenses.

7. Quick Button

Press this button to access customized preset settings. We will talk about these in Step 2.

8. POWER Button

Press this button to turn your camera ON.
Once your camera is powered on, press this button to put the Touch Screen to sleep or to wake it up.
Hold down this button to turn your camera OFF.

9. Speaker

This speaker plays audio when watching videos on the Touch Screen.

10. USB Port Door

The USB port is for charging and transferring files. The rubber gasket on the door keeps water out of your camera, so make sure it is always closed securely.

11. Battery and MicroSD Card Port

Open this door to install or replace the rechargeable battery and to insert or remove your microSD card. Step-by-step instructions for transferring your files are provided in Step 5- Creation. This door is removable. The rubber gasket on the door keeps water out of your camera, so make sure it is also always closed securely.

Microphones

Four waterproof microphones record stereo and 360 sound.

NAVIGATING THE INSTA360 X3

Let's turn your camera on and learn to navigate your new camera so you can tap into all of the fun features of this powerful 360 camera.

The simplest way to get around the Insta360 X3 is to use the camera's integrated Touch Screen on the back of the camera. All of the modes, settings and setup options are available through the Touch Screen.

There are lots of setting options on the Insta360 X3 (which we will get into in the next step), but after a little practice, you will be able to navigate your camera with ease. First, let's take a closer look at navigating using the Touch Screen.

If you want to take a moment to practice getting around, follow the steps to take a short tour.

1. **Press the Power/Mode Button on the side of the camera to turn your camera on**. It will take a few seconds for your camera to power on.

2. The first time you turn the camera on, the X3 will power on in **360 Mode**, which means that it will record through both of the lenses for spherical 360 videos and photos. The X3 also offers a Single Lens Mode and can easily be switched between the two. To switch between 360 and Single Lens Modes, just **Tap the camera icon in the bottom left corner of the Touch Screen OR Swipe Left or Right from the center of the screen.** The option to switch between 360 and Single Lens Mode is at the top of the screen. For now, let's stay in 360 Mode.

> **TIP:** If the Touch Screen goes to sleep on you, just press the Power Button to reawaken the screen.

3. In both SINGLE LENS and 360 Modes, the X3 has **a variety of recording modes.** After tapping on the icon in the bottom left corner or swiping left or right from the center of the screen, the different modes are indicated by the **icons in the middle of the screen.** You can Swipe Left or Right to scroll through the different mode options, which offer options for recording videos, photos, and time lapses. You will learn how to use these modes in Step 2. After selecting a mode, the icon at the bottom left corner of the screen indicates your current mode.

4. Next, **Tap the curved arrow icon** at the bottom right of the touch screen. This switches lenses in Single Lens Mode. In 360, this just switches the preview lens since the camera records using both lenses. (You can also use the Lens Button to scroll through lenses.)

5. Now, at the **bottom middle of the Touch Screen, Tap the settings box to adjust a few key settings such as Resolution**. Once you select a setting, it returns to the main screen.

6. Swipe Left (from the right side of the screen) to access the remainder of the settings for that mode. **We will cover these settings in the next step so don't change any of these now.**

7. **Next, Hold Down on the middle of the screen** and all of the icons will be hidden for composing your shots more clearly. Tap the "x" and they reappear.

8. To begin recording after selecting a Mode you would **press the Shutter Button.**

TIP: Using Buttons to Change Modes. When the Touch Screen is inaccessible, there are two ways to switch between modes using the buttons. This first option is to enable Dive Case Mode using the Shortcut Icon on the Touch Screen. Dive Case Mode disables the Touch Screen. You can then scroll through the Modes by pressing the Power Button. To select your desired Mode, press the Shutter Button. To switch lenses between 360, single front lens, and single rear lens, press the lens button. To disable Dive Case Mode, unlock the Touch Screen.

The second option is to press the Quick Button on the right side of your camera to quickly switch between customized presets. Press the Quick Button until you reach your desired Preset. Then press the Shutter Button to select and return to the recording screen.

TIP: The **Shortcut Menu** (including the general camera settings) is accessed by Swiping Down from the top of the screen. We will go over those in Step 2. The recorded media is accessed by Swiping Right from the left side of the screen.

QUICKCAPTURE (ONE BUTTON CONTROL)

QuickCapture enables you to power on your camera straight into recording without any standby time. When recording is finished, the camera automatically turns back off.

To use QuickCapture, you need to first enable the feature when your camera is turned on. Swipe down from the top of the screen and tap the "Rocket" icon. Now QuickCapture is enabled.

When your camera is turned off and you want to get straight into recording, utilize QuickCapture for quick access to video recording with the push of a button. **To begin recording video, Press the Shutter Button** and your camera will power on and begin recording 360 video. Once you **press the Shutter Button again, your camera stops recording and turns off.**

QuikCapture conserves battery life that would be used during standby time. Note that there is an 8-10 second startup time when using QuikCapture.

THE TOUCH SCREEN

The Touch Screen is a very convenient feature of the X3. The Touch Screen gives you the ability to preview and compose your shots, as well as see a Live View of how setting changes affect your videos and photos.

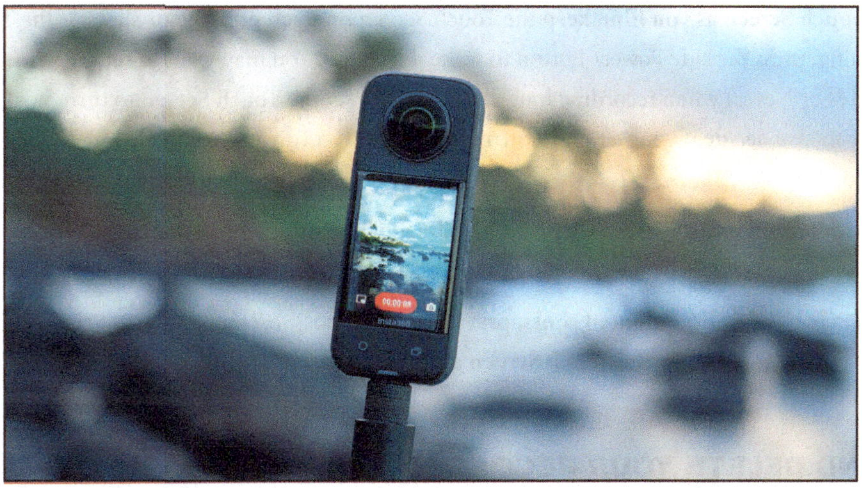

You've already learned how to navigate the Touch Screen to change modes and basic settings, but **the following section provides more information about other useful ways to tap into the convenience of the Touch Screen.**

The Touch Screen serves three primary functions: to **change settings,** as a **viewfinder**, and to **view your recorded photos and videos**:

TO SELECT AND CHANGE SETTINGS

As you learned previously, all modes and settings can be changed using the Touch Screen.

The Touch Screen tells you the following information about your camera settings, modes and status:

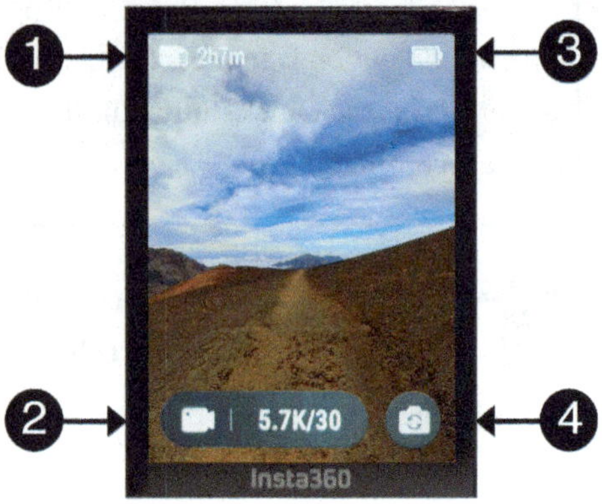

1. Time/Storage/Files

In Video Mode, this displays the remaining number of minutes you can record on your memory card at the current video resolution. (This time will change when you change settings because different file sizes require different amounts of memory.)

In Photo Modes, this tells you how many photos you have remaining at the current setting.

2. Camera Mode

This icon indicates which recording mode you are currently using and the key settings for the mode, such as Resolution and Frame Rate for Video.

3. Battery Life

This displays the remaining battery life.

4. Switch Lenses

Tap this icon to switch between the front and back lens. In Single Lens Mode, the selected lens will be the only one recording. In 360 Mode, this only affects the preview since both lenses will used to capture the image.

• **To hide the settings info,** Long Press on the Screen. Hiding the info while you record allows you to see your shots more clearly for better composition. To show settings info on the preview screen, Tap the "x" on the screen.

• **To escape** out of a Setting Menu, either 1) select a Setting, or 2) press the Back Arrow.

• Swipe Down from the top of the screen to **access the Shortcut Menu** for more general camera settings. (You will learn about these at the end of Step 2.)

AS A VIEWFINDER

The Touch Screen provides an easy way to **set up and compose your shots.** Use the Touch Screen to preview how your shots will look with the selected settings. For a clear screen after selecting your settings, Hold Down on the Touch Screen to hide the icons.

When looking at the Touch Screen as you film, keep the Touch Screen on to compose your shots. If the Touch Screen goes to sleep while you are filming, press the side Power/ Button to wake it back up. You may want to set the Screen Saver for a longer period (such as 3 minutes or Never) when recording long clips. You can set the Touch Screen to turn off after 30 seconds, 1 or 3 minutes or Never. You can also decide if you want it to turn off 'During Recording'. The Touch Screen settings can be set in the Preferences Menu>Auto Sleep.

While recording 360 video, you can adjust the perspective by dragging your fingers on the Touch Screen. Tap the icon in the bottom left corner for MultiView to see the view looking both directions simultaneously.

When using the Touch Screen to **set up mounted shots,** preview the composition on the Touch Screen. Once the shot is set up and your mounting position is secured, **let the Touch Screen go to sleep** to conserve battery life.

TO PREVIEW AND DELETE YOUR RECORDED PHOTOS AND VIDEOS

After a session, use the Touch Screen to view your videos and photos. One of the great things about the Insta360 X3 is the instant gratification of seeing your photos and videos right away on the Touch Screen. The built-in screen lets you view both your 360 and single lens footage immediately.

• **Swipe Right** from the left side of the screen to open your media. The diagram shows Touch Screen items for playback (icons will vary slightly depending on the type of media you are viewing).

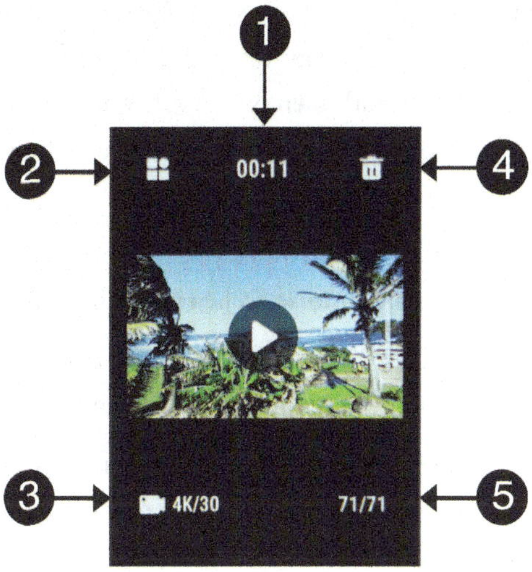

1. The **Time Counter** shows a running count of your play back time.
2. Select the **Grid Icon** to view thumbnails of your media.
3. This Mode Icon shows the **Recording Mode used and Settings** for that file.
4. Tap the **Delete** Icon to delete the current file.
5. Displays the **total number of files** on the microSD card.

• **Swipe Up or Down** from the middle of the screen to view more videos or photos.

• **After tapping the Grid,** use the Gallery view to select files one by one, or Tap the Check box in the top right corner to select all files. Delete the selected files by tapping the Trash Can icon.

• **Push play/pause in the middle of the screen** to watch the video, view a single shot photo or view a sequence of photos.

• **Drag on the screen** to adjust the view of a video or photo.

• **Swipe Left from the right side of the screen** to escape out of the media tray.

VOICE CONTROL

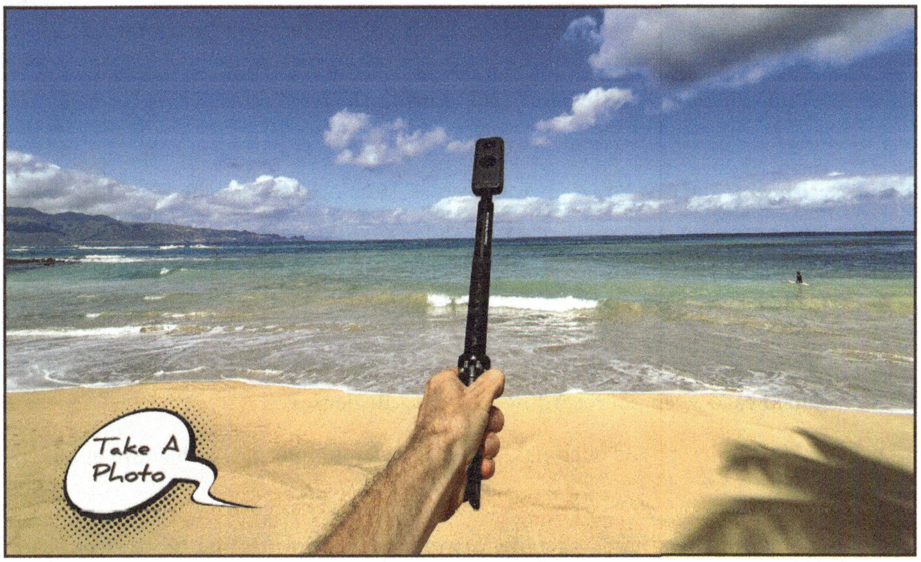

At first, Voice Control seems a bit like a gimmick, but as you start to use it, you will realize that it's actually **very, very useful when using a 360 camera**. Voice Control gives you easy hands-free access to start and stop recording with your Insta360 X3. Since your camera records a full 360 sphere in 360 Mode, Voice Control gives you the freedom to move away from the lens before recording. Voice Control is also useful to control your camera when you are too far away to reach the Shutter Button, such as on the end of a long extension pole. You will learn more about the importance of this in later chapters.

To activate Voice Control, Pull Down the top tray on the Touch Screen. Tap the Settings. Under Voice Control, turn Voice Control On. The preferred Voice Control language can also be selected in this menu.

You can't do everything with Voice Control, such as switching between 360 and Single Lens Modes, but there is a lot you can control. **Here is a list of the commands you can use with your Insta360 X3:**

To Start and Stop Recording or Taking Photos

"Start Recording" (starts recording video)
"Stop Recording" (stops recording video)
"Mark That" (mark memorable video moments for the AI editing)
"Take a Photo" (if your camera is not recording a video)
"Shut Down Camera" (to power off)

Of course, as with anything, there are optimal times to use Voice Control. For best results, use Voice Control when there is **minimal ambient noise.** When recording video, you need to **stop recording before issuing a new command.**

Also, keep your X3 **within a few feet** to make sure it can hear you, especially if there is ambient noise, such as the ocean or wind. Speak clearly and say the exact command.

Also, Voice Control will not work when your camera is underwater, or even sometimes after you have been in the water (if there is sand or water covering the mic opening). Make sure to shake your camera or blow any water out of the mic openings after going in the water.

TAKING CARE OF THE INSTA360 X3

Before we get too deep into using your new X3 camera, there are a few things to remember to keep your camera in top condition.

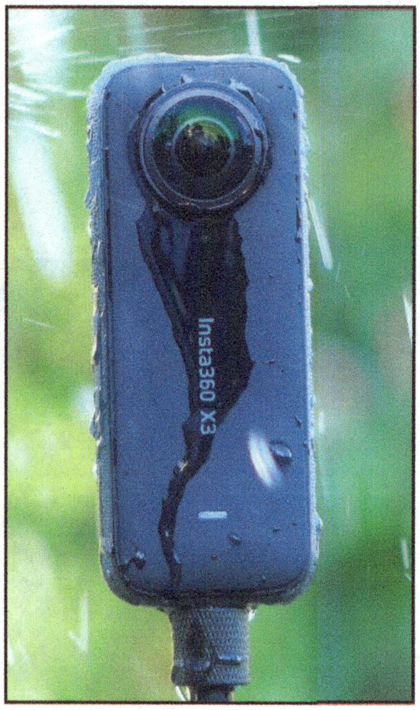

These simple steps will keep your camera working like new for years:

• Keep the side USB door and the battery closed with the tabs completely closed (with no orange visible) **before getting your camera wet.** The X3 is waterproof to 33' (10m), but only when the door and the battery compartment are closed completely.

• **Rinse off your camera and mounts with fresh water** after using them in the ocean or getting them dirty. Gently dry your camera with a soft cloth and **blow off the residual water around the door** before opening it.

• **Avoid touching the lenses with your fingers.** You don't want grease, sunblock, or scratches on the lenses. Make sure the lenses are always clean! This is your camera's window to the world. The Insta360 X3 **works great for clear, in-focus shots** as long as the lenses are clean. The lenses on the X3 are not replaceable, making it extra important to prevent cracks or scratches.

• **Keep the camera in the included case** when not recording to protect the lenses.

• Try to keep sand and dirt out of the microphone openings. You don't want to interfere with audio quality by blocking these openings.

• The Insta360 X3 lenses can't handle too many hard hits so **if you want the extra protection for the lenses, use the Sticky Lens Guards** for extreme sports. If you use the Sticky Lens Guards, make sure to set the stitching for Lens Guards in the Shortcut Menu.

WIFI / BLUETOOTH

The Insta360 X3 uses a combination of 2.4 and 5GHz WiFi and Bluetooth (BLE 5.0) to connect to external devices, such as a remote or a phone/tablet. Your camera emits its own signal, so you don't need to be within range of any WiFi signals to use your camera's WiFi features. There is no indicator light or icon to tell you if your WiFi signal is on.

To access the WiFi dialog, pull the top tray down on the Touch Screen. In Settings, Tap WiFi Settings to see your WiFi settings and switch between Auto or Always On.

USING THE INSTA360 APP

The Insta360 App works on your tablet or smartphone and is available for free from the App Store (iOS) or Google Play (Android). The App is optional, but the benefits are many.

In addition to giving you remote access to your camera's Shutter Button, modes and settings, the Insta360 App allows you to **view the action as you record** videos and photos when you can't see your camera's Touch Screen. This is extremely useful for composing your mounted shots. You will see the benefit of this technology when you learn to set up your shots in Step 3- Mounting.

TIP: **Live View** (viewing what you are filming while your camera is recording) compatibility depends on your device, but if you see a "Preview Not Available" message once you start recording, you are filming in a resolution that is not compatible with your device.

The Insta360 App is also the **easiest way to transfer photos and videos** from the X3 to your phone or tablet for sharing with friends, family or on social media. The Insta360 App also gives you the best tools for editing 360 photos and videos, which we will learn more about in Step 5- Creation.

You can also **update your camera's firmware wirelessly through the App.** Updating your camera ensures that your camera is equipped with the most up-to-date features and settings. The App will notify you when an update is available.

TO CONNECT THE INSTA360 X3 TO THE INSTA360 APP:

Download the Insta360 App from your app store. The Insta360 App is free so make sure you don't accidentally download an app that looks similar but charges a fee. There are no additional upgrade options for the Insta360 App.

After you install and open the App, tap the yellow camera icon in the bottom center of the screen. The App will walk you through the setup.

When reconnecting to the App, power your camera on first. Then, you should just be able to open the App and tap on the camera icon to connect

If the camera is still not found, make sure your camera's WiFi is turned on in Settings>WiFi Settings and the phone's WiFi connection is connected to your camera.

UNDERSTANDING THE APP

Once your camera is connected wirelessly to the App, the Insta360 App gives you 2 options: 1) Control Your X3, which we will explore more here. 2) View Media, where you can view, transfer, and edit the media on your camera's microSD card. We will dive deeper into that in Step 5, when you learn to edit your media.

CONTROL YOUR INSTA360

When you connect your camera to the App using the yellow camera icon, the following icons give you remote access to the full range of settings and modes on the Insta360 X3 (Don't worry, we will talk about what all these settings mean in Step 2):

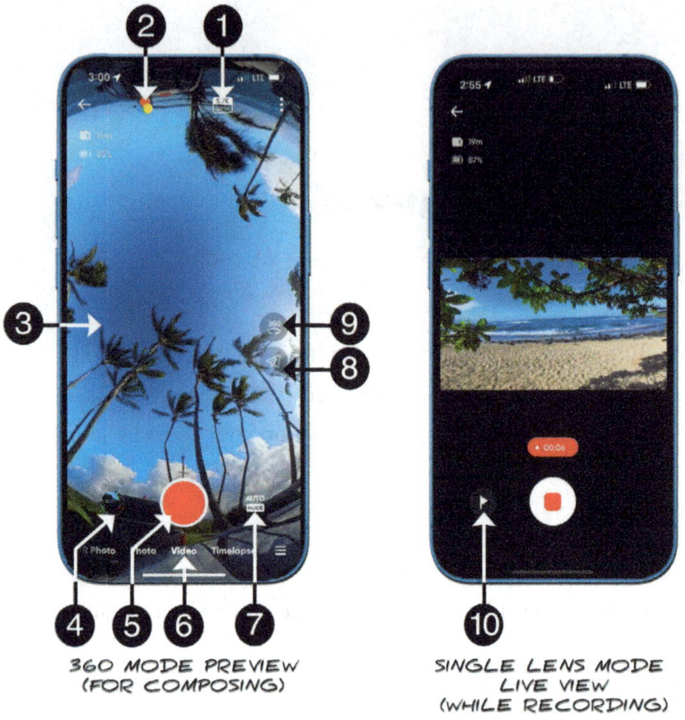

1. Shooting Parameters (Key Settings)

Tap this icon to change the video resolution and frame rate In Video Modes, or, to change resolution, aspect ratio and delay timer settings in Photo Modes. For some Photo Modes, a Pure Shot option will also be available.

2. Color Profile

Switch between the three color options. (Only available in some video modes).

3. Preview Screen

This screen shows a preview of your camera's view. In 360 Mode, the screen shows a full frame view on the App. As you rotate and move your phone or tablet, you can look around at the camera's full 360 view. You can also zoom in or out by pinching with two fingers or dragging up and down to adjust the composition. All of these adjustments are only to preview your shots. This full 360 view is useful for composing the angle and distance from your subject, but the camera will still record the full 360 sphere, which can be edited later.

In Single Lens Modes, the preview shows the actual frame that will be recorded.

4. Insta360 Media

Tap this icon to view the videos and photos that are currently on your camera's memory card. You can view, trim, download, edit, delete and share these files.

5. Shutter Button

Press the Circle Button on your selected mode to begin recording video or taking photos.

6. Modes

Swipe left or right to change between modes. Press the Menu icon on the right for more available modes.

7. More Settings

More settings for the current mode, such as ISO and White Balance are located here. In Step 2- Settings, you will learn which settings to use.

8. Fly Through Mode

Tap this icon to enter Fly Through Mode. We will talk more about using Fly Through Mode to create pass through shots in Step 4- Camera Movement.

9. 360 or Single Lens Mode & Switch Lenses (Single Lens Mode Only)

Tap this icon to switch between 360 and Single Lens recording. In Single Lens Mode, use the curved arrow icon below to select between the front and back lenses.

10. Mark (displayed only when recording video)

Press this button while recording video to add a Mark of your best moments. These Marks help the Insta360 App determine which moments to highlight when editing.

After filming, you can view, edit, delete, and download the content from your camera's memory card. When looking at your media, you may see a low resolution preview. This is not the actual quality of the footage. You can however edit these files and export even when the media is still on your camera's microSD card. You will learn more about mobile editing using the Insta360 App in Step 5.

TIP: Insta360's Desktop App. We will learn about desktop editing in Step 5, but if you want to download Insta360's desktop editing app called Insta360 Studio, this app can be downloaded from Insta360's website. You will need to provide your camera's serial number, which is located in the Shortcut Menu>Settings>Camera Info. We will go through Insta360's editing tools in Step 5.

INSTA360 CARE

Just so you are aware of it, for an additional fee, Insta360 offers accidental damage protection called Insta360 Care. Insta360 Care gives you one free repair or replacement for your Insta360 camera, so you can relax and enjoy the adventure.

If you break your camera, which is a possibility if you are out adventuring, you can replace your camera once for free.

TIP: Offloading Footage To A Computer. If you want to skip ahead and offload your footage to a computer, there are a few ways to transfer your recorded files from your camera's microSD card (where they are stored) to a computer. Step-by-step instructions for transferring and organizing your files are provided at the beginning of Step 5.

NOW THAT YOU'VE GOT YOUR CAMERA UP AND RUNNING,

LET'S GO TO STEP 2 TO LEARN ABOUT THE MODES AND SETTINGS.

STEP TWO

THE SETTINGS

Choose Your Settings to Get the Shot

When it comes to capturing unforgettable moments with the Insta360® X3, the options are endless. The Insta360 X3 is a powerful little 360 camera, packed with options for creating mend-bending videos and photos. Because the X3 is such a powerful recording tool, there are lots of setting options to choose from! The **biggest factor to recording those unforgettable moments is setting up your shots correctly.** To a new 360 camera user who may not yet realize the full potential of the footage you are about to capture, the setting options may be a bit hard to filter through. In this step, we will break it down to give you the simplified guide to the Insta360 X3's settings and modes so you can learn which recording options to use and when.

As you begin to unlock the power of your new 360 camera, it's important to realize that everyone has different tastes and interests. Some of you may love editing high quality videos on a computer, while others may prefer mobile editing for social media. Some of you may love the spherical 360 look, and some of you may prefer a more traditional look for your photos and videos. Whatever your pleasure, **we will give you the knowledge you need to customize your Insta360 X3 to your tastes** so you can take your footage to the next level of amazing!

In both 360 and Single Lens Mode, there are **three main types of media to record; Video, Photo and Time Lapse.** Each type of media functions slightly differently so we will approach each type of media as its own group. The fourth section in this step highlights some of the general camera settings available in the Shortcut menu to customize your filming experience.

Because there are a lot of setting options available on the X3, explaining the settings is the most technical step. Once you get through this first step, the rest will be a breeze. Just try to get a basic understanding of the main settings and the rest will fall into place. So, let's get into the art of capturing those magic moments.

NOTE: Make sure your camera is running firmware version v1.0.21 or higher so that the following information matches your camera's settings.

360 VS SINGLE LENS MODE

As we touched on briefly in Step 1, the Insta360 X3 offers two styles of recording photos, videos or time lapses- 360 or Single Lens Mode. The first, called **360 Mode, records a full 360 degree sphere**. The second mode, called **Single Lens Mode, records a more traditional rectangular frame.** Let's learn a bit more about each one.

When 360 is selected at the top of the Touch Screen, the camera records with both lenses capturing a full 360 degree video or photo. Because the camera records the full 360 view, this gives you lots of freedom to edit your footage later, either keeping it in 360 or editing the footage into a more traditional rectangular format.

In 360 mode, the camera actually records two separate videos. Each one is slightly more than a full hemisphere. These two videos are then "stitched" together, creating a full 360 sphere photo or video. The overlapping areas disappear without losing any of the scene which opens up some amazing perspectives and mounting opportunities.

In **Single Lens Mode, the X3 uses only one of the lenses to capture a traditional rectangular image.** Even though it doesn't record the full 360 view, Single Lens Mode uses the super wide angle lens to produce amazingly smooth videos, immersive photos, and some other unbelievable features.

You will learn more about how to take advantage of 360 and Single Lens Modes as we move through this book.

Recommended Settings

As we make our way through this section, we will give you the best recommended settings for each mode. The settings can always be changed later as your personal preferences develop, but these recommended settings are a great starting point for high quality photos and videos.

RECORDING VIDEOS

This section helps you understand the different modes and settings when using the Insta360 X3 to record videos. Let's go through the following three steps to help you get the videos you dreamed of when you bought your X3!

1. First off, **let's learn about the available video modes so you can understand how each one works.** We will also provide the recommended settings to start with for each mode.

2. Next, we will walk you through the settings to help you **understand how each setting affects your videos.**

3. Then, after you understand more about the settings, we will **give you a few more tips for selecting the best settings and modes** to get the most out of your Insta360 X3.

Just as a reminder, to change modes, swipe sideways from the middle of the screen or press the mode icon at the bottom of the screen.

There are two icons used to access and select settings for the current mode:

On the main recording screen, Tap the numbers next to the Mode Icon (5.7k/30 for example), to change the main settings for that mode, such as Resolution, Frame Rate, and Aspect Ratio (for Single Lens Videos only).

The rest of the settings for that mode can be accessed by Swiping Left from the right side of the screen. This opens up additional Settings such as Color Profiles, White Balance, and more that you will learn about in this step.

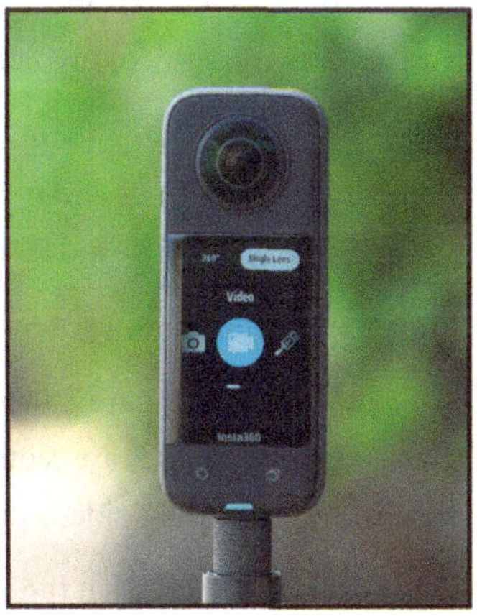

1. VIDEO MODES (PRESETS)

When recording video with the X3, first decide if you want to record 360 Video or Single Lens Video. We will look at the options for each style of filming:

360 VIDEO OPTIONS

When 360 Mode is selected from the Mode Menu, your camera will record using both lenses. Four of the 360 modes are designed for recording video, or a variation of video- 360 Video, Active HDR Video, Bullet Time and Loop. Let's look at the best uses for each of these:

VIDEO (360)- The Highest Quality Full 360 Videos with tons of potential for creative editing!

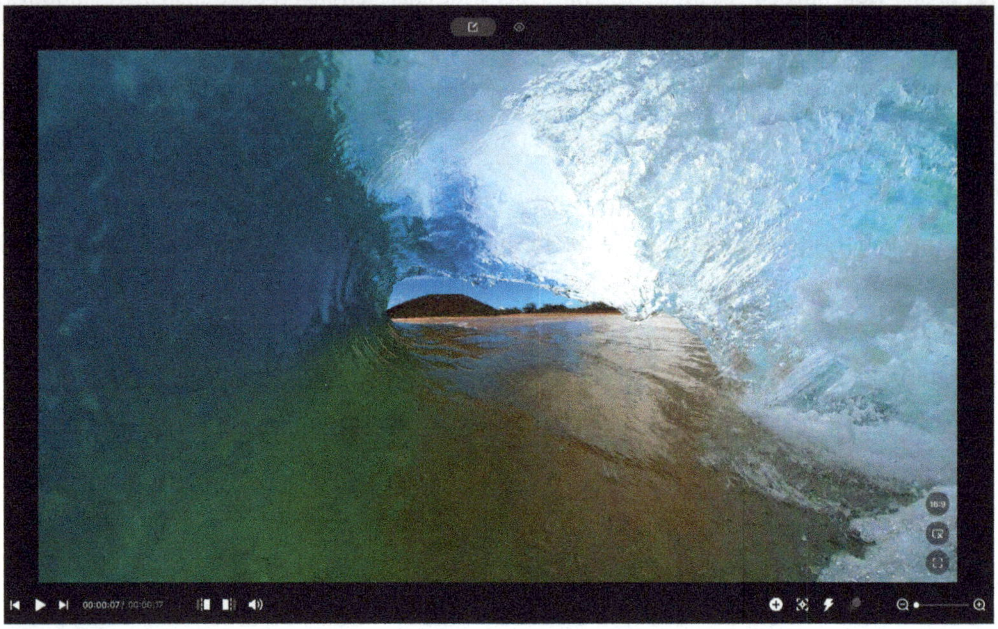

360 Video records a full 360 spherical video, offering the most editing flexibility. 360 Video recorded in Video Mode with the X3 mounted on the Insta360 Invisible Pole.

Use this mode to record **stabilized high quality 360 videos.** In 360 Video Mode, your camera uses both the front and back lenses to **record a full 360 degree sphere.** This is the mode to use as your **go-to for recording normal 360 video a variety of lighting situations.** Because you will be recording a full sphere, 360 video has so many creative possibilities when you go to edit your videos.

Besides the fact that this mode records full 360 video, the camera operates like any other video camera. Press the Shutter Button to start recording video. Press the Shutter Button again to stop recording.

360 video doesn't offer a lot of extra frilly features because the magic mostly happens when you go to edit your videos. There are some advanced settings and filming tips that you will learn to improve the initial raw video out of the camera. We will cover these advanced settings as we move through this step, as well as mounting and filming tips in the next two steps.

Even though the editing potential for 360 video is almost limitless, starting out with high quality video is key.

• 360 video files from the X3 end in the filename ".insv". When you record a 360 video, the X3 will create two files, one from each of the lenses. Both files are required for editing. The camera will also create a .lrv file. The LRV is a low resolution version, which is used for previewing videos using the Insta360 App.

We are going to learn what the settings mean next, but you can start with the following settings as your go-to for most video shots.

The Best 360 Video Settings:

• To start with, the best setting for recording 360 video in **5.7k @ 30 FPS** in Auto Mode with Vivid Color.

• To record 360 video **for slow motion**, switch over the RES | FPS setting to 4k at 60 FPS. Or you can use the super slow motion editing trick we will show you in Step 5 if you recorded in 5.7k at 30 FPS.

• Videos recorded in 360 Video Mode are **automatically stabilized** and there is no option to turn this off (not that you would want to).

ACTIVE HDR VIDEO (360)- 360 Videos recorded with a Higher Dynamic Range.

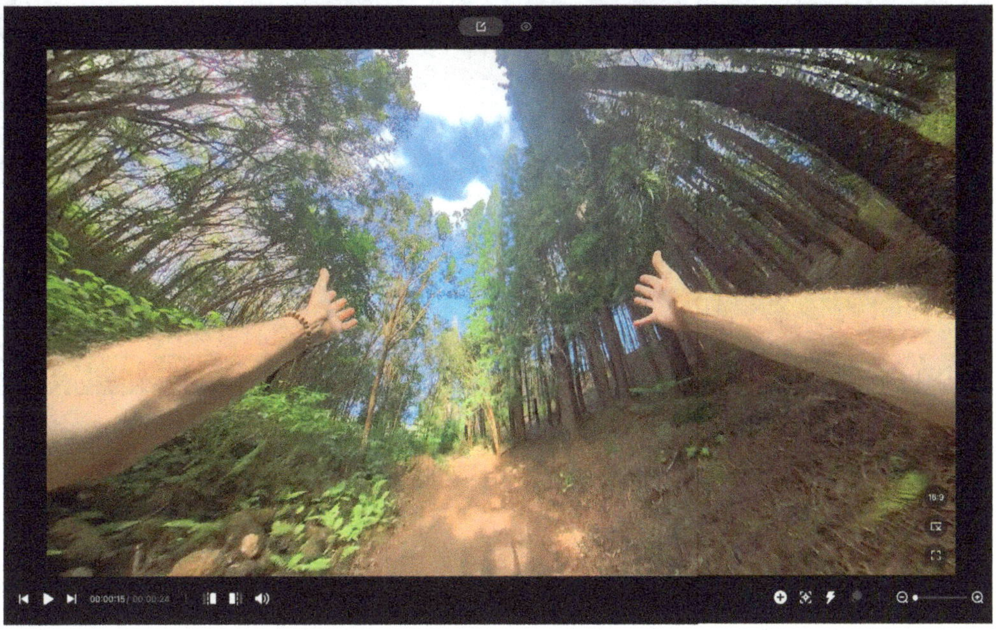

Active HDR Video balances out bright and dark areas for a more even tone throughout the video. However, you can see some jittery effects because of the slightly low light scene in the forest. This can ruin your videos, so only use HDR Video in full bright light. 360 Video recorded in Video Mode with the X3 mounted on a Bite Mount.

Videos recorded using the Active HDR Mode have less contrast in bright and dark areas, giving the video **a more balanced range of tones**. Active HDR should **only be used in bright light scenes** because the HDR processing will produce undesirable effects in any low light scenes.

The best time to use Active HDR is in tricky lighting conditions, such as midday lighting, when the light is harsh.

Even though HDR videos produce a nice balance in low light scenes (at sunset for example), this is not the best time to use Active HDR. If you do want HDR videos in low light scenes, set your camera down in a stationary position to prevent jittery footage.

THE BEST ACTIVE HDR 360 VIDEO SETTING:

- The best setting for recording HDR video in 5.7k @ 30 FPS with Vivid Color.

BULLET TIME (360)- Recorded using both lenses to create a rotating 360 effect in slow motion. Matrix-Style.

Bullet Time creates an amazing effect where **the camera rotates around you (or your subject) in super slow motion.** When Bullet Time Mode is selected, the camera uses both lenses to record, but the resulting video is **a traditional flat video.** This is a great mode to use to highlight yourself in any scene, and those who watch it will be awed by your "special effects" editing skills. But really, it's all in the mode and Insta360's AI.

To use Bullet Time, attach your camera to the Bullet Time handle or the Bullet Time cord. Select the Bullet Time Mode. When you are ready to start recording, press the Shutter Button and start swinging your camera. The key to success with this mode is to get your camera swinging around you in a steady circle at the right speed, which is about 1 second per rotation.

You can either stand still, or for more movement in your shots, walk/run forward while swinging the camera.

The camera automatically recognizes that Bullet Time Mode was used and processes the video in-camera. You can select the video aspect ratio (we will talk about that in this step) for your final video. There it is. Magic in the swing of your fingertips!

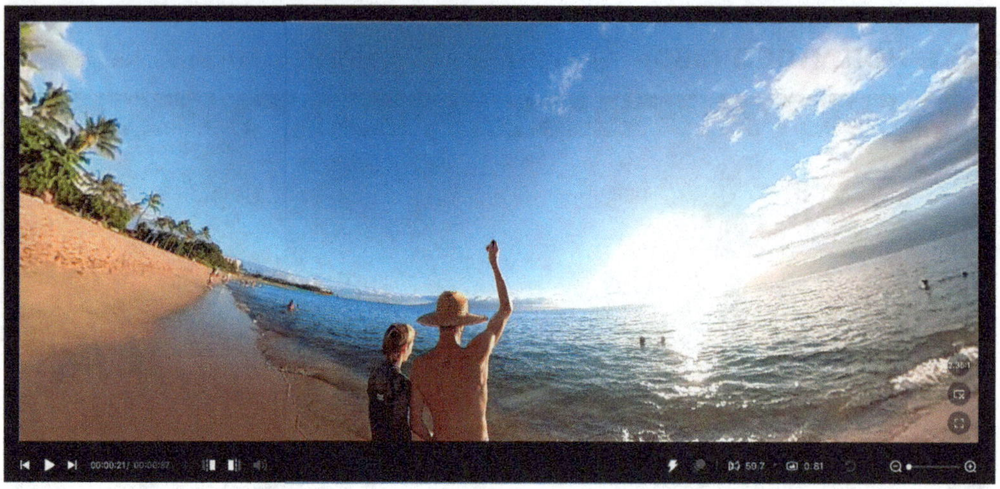

Bullet Time produces an eye-catching video that revolves around the subject in super slow motion for a captivating addition to any video edit.

The **best setting** for recording Bullet Time videos in 4k @ 120 FPS with Vivid Color. The 120 FPS setting produces amazing super slow motion for your Bullet Time shots.

LOOP RECORDING (360)- 360 Looping Video. Great for Dash Cam Setups.

Loop recording allows you to **record a continuous 360 video for a set duration of time** until it begins to record over itself. The primary reason to use this mode is to use your X3 as a dashcam. Once something happens that you want to keep a recording of, press the Shutter Button and the camera will save the most recent period of time. The duration options range from 1 minute up to 30 minutes. If you are recording for an extended period of time, you will need to connect your camera to a USB power source. This is easy to do, especially if your camera is mounted inside of your vehicle.

SINGLE LENS VIDEO OPTIONS

Single Lens options produce a traditional flat video, but there are also some really useful modes in this set of options. Three of the presets in this mode are designed for recording video- Video Mode, Me Mode and Loop Mode. Let's look at each one, then we will look at some of the setting options that are unique to the Single Lens Video Modes:

VIDEO (SINGLE LENS) - Traditional Rectangular Format Video with the widest range of settings and the most compatibility with devices.

In Single Lens recording, the Video Mode option is used to **record a more traditional rectangular frame.** The X3 can record high resolution videos up to 4k using either of the lenses. With the high resolution available on the X3, the settings in this mode more closely match the traditional video settings found on most action cameras.

In Single Lens Video Mode, the camera only records what is shown on the Touch Screen.

Single Lens Video Mode is a useful option **when you want videos ready to go**, straight out of the camera without any editing required. If you have a specific vision for your video and can capture it with a single lens, this is a great mode to use.

To record videos in Single Lens Video Mode if you are already in 360 Video Mode, simply use the "Switch Lens" icon to select your lens and start recording. Or Tap the Mode icon on the bottom left corner of the screen to enter the Mode Menu and select Single Lens>Video Mode.

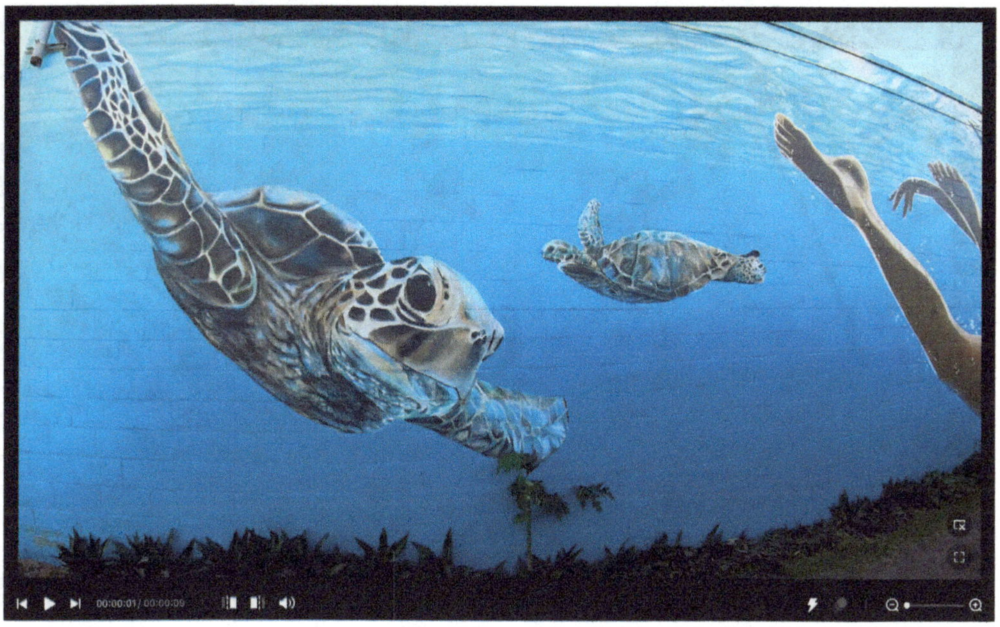

Single Lens Video Mode records a traditional rectangular video, which is great when you don't want to reframe your videos. Video recorded in Single Lens Video Mode with the X3 handheld on the Insta360 Invisible Selfie Stick.

Best Single Lens Video Settings:

- The **best regular speed video setting** is 4k @ 30 FPS in Auto Mode with Vivid Color for regular speed playback. This is the highest resolution available.
- To record **single lens videos for slow motion,** switch over the RES | FPS setting to 3.6k at 60 FPS for up to 2.5x slow motion. Or you can use the super slow motion editing trick we will show you in Step 5 if you recorded in 5.7k at 30 FPS.
- Videos recorded in Single Video Mode are automatically stabilized and there is no option to turn this off (not that you would want to).
- We will talk about lenses later in this step. Select a lens that suits your shot prior to recording since Single Lens Mode videos don't have the option to change lenses in the App like 360 videos allow.

LOOP (SINGLE LENS) - Record looping Videos at a designated length.

This mode is similar to Loop Mode in 360 Mode, but instead of recording a 360 video, this mode records a looping traditional video out of just one of the two lenses.

The **best setting** for recording looping videos is 4k @ 30 FPS since you probably don't need slow motion. If you want the ability to play your videos back in slow motion, choose 3.6k @ 60 FPS.

ME MODE (SINGLE LENS) - Film selfie videos using the Invisible Selfie Stick. The pole will magically disappear, giving you the magic of 360 in a traditional video frame.

When filming using Me Mode, **hold your camera straight out in front of you close to waist level.** The camera will frame you in the shot, but the pole or Invisible Selfie Stick will not be visible. This mode is a great alternative to 360 Video Mode when you want less editing but still want the invisible pole effect of a 360 shot. The resulting video will be a traditional flat video.

The **best setting** for Me Mode is 1080 @ 60 FPS using the Max View Lens for the widest perspective. For less distortion, use Narrow, but you will need to extend the selfie stick further from you.

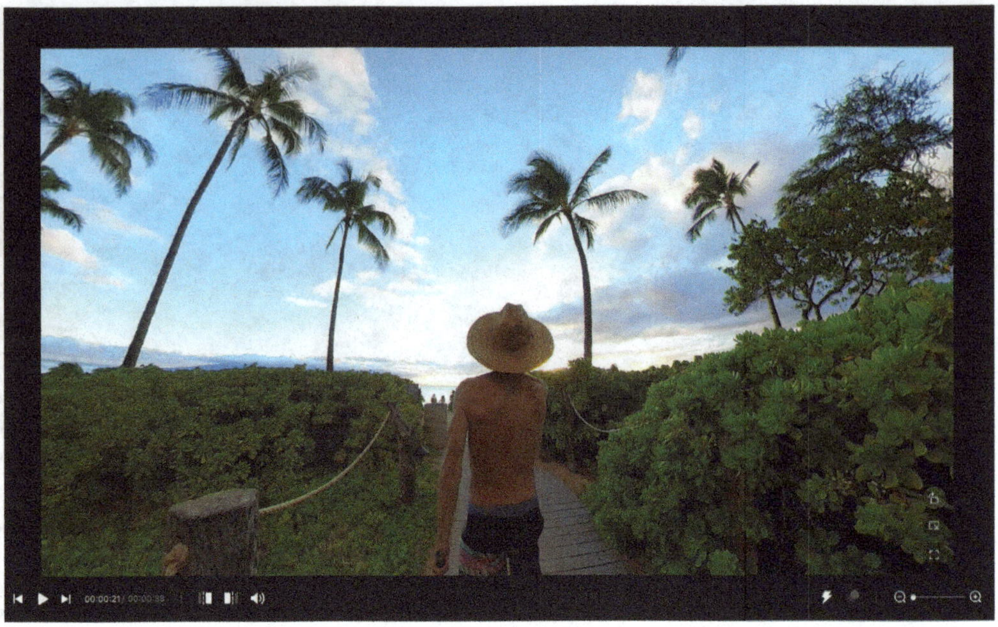

Me Mode records a traditional rectangular video without the selfie stick visible in the shot. Video recorded in Me Mode with the X3 on the 114cm Invisible Selfie Stick.

2. VIDEO SETTINGS

Now that we've covered the video modes available, let's look deeper into the settings available.

THE BIG 3 VIDEO SETTINGS

Now that we have covered the video modes, we will cover the big three video settings. These first three setting options are the most important when it comes to selecting the right settings for your videos. They determine the quality, playback speed and framing of your video. These are basically **the physical attributes of the video you are recording.**

Selecting the best video settings for your moment will help you achieve the shot you want and give you the freedom to edit your video how your mind envisions. Whether your camera is mounted near or far, you are filming fast action or scenery, full sun or low light, the settings you choose here will produce high-quality footage to work with when you edit your videos. Understanding how these settings affect your videos will help you take full advantage of the best X3 settings for a specific scene or activity.

These three settings are adjusted using the main setting options next to the Mode Icon. ***Note****: Aspect Ratio is only available in Single Lens Modes since the Aspect Ratio can be selected when reframing your 360 videos to a traditional video format.*

1) RESOLUTION

RES stands for resolution and this determines the actual size of your video image. Depending on how you plan to share your video, this can be one of the most important settings.

When choosing settings, you will see the file size displayed like this: 5.7k-30, which is also sometimes written like this: 5.7k @ 30FPS. The first number before the dash ("5.7k" in 5.7k-30) tells you **the size of the video image**. So, why does resolution matter?

In simple terms, a video "image" is made up of a bunch of small dots, called pixels. A video comprised of many pixels creates a larger, more detailed image.

The resolution name (5.7k, 4k, 3k, or 1080p) defines how many lines of pixels exist within the video frame. **The higher the number, the better the quality**. As you can see in the chart below, a 5.7k video is much larger than a 3k video and produces more detailed video.

In 5.7k, 4k, and 3k, etc., the names refer to the width of the image in pixels. For example, a 5.7k video is about 5600 pixels wide (after stitching).

In 1080p, the number refers to the vertical height of the image being recorded. For example, in 1080p, the image is 1,080 pixels high. (The "p" after the number stands for "progressive scan," which, without getting too technical, creates smoother, more detailed video).

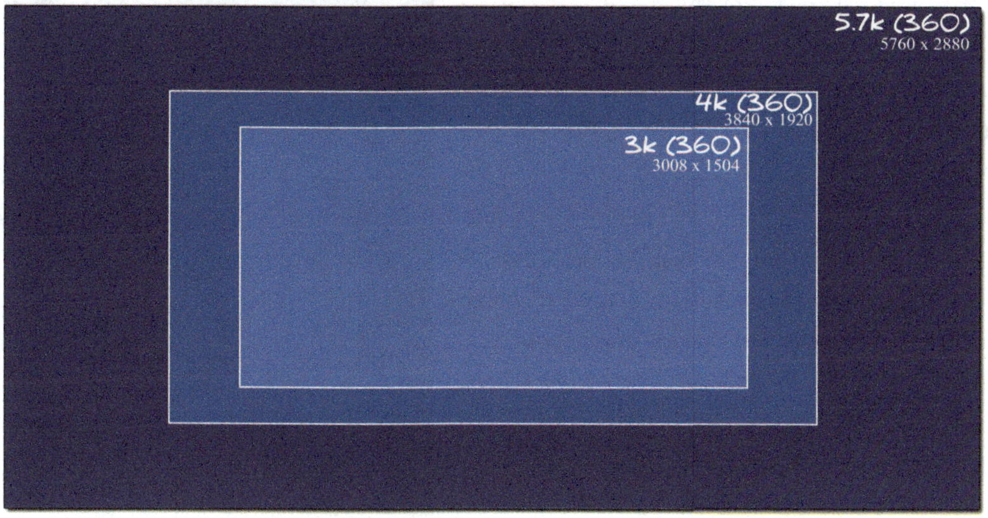

This chart shows the relative pixel dimensions (width x height) of the 360 video resolutions available on the Insta360 X3.

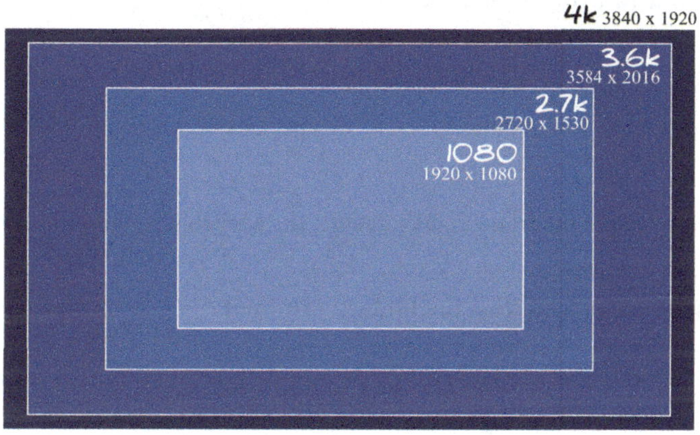

This chart shows the relative pixel dimensions (width x height) of the SINGLE LENS video resolutions available on the Insta360 X3.

The Lowdown

360 Video Resolutions

• **5.7k 360** video is used for regular speed playback. This is the highest resolution video available on the X3 and should be your first choice. Reframing 5.7k video typically results in a traditional 1080p video, but you can export to 4k when using Insta360 Studio.

• **4k 360** video is used for 2.5x slow motion. You will only want to reframe this resolution to 1080p, but depending on the quality of the original video, you may be able to get away with Ultra Wide Angle shots in 4k.

• **3k 360** video is a much lower resolution, but it offers higher frame rates (coming up next) so you can play back the video in super slow motion. Reframed 3k 360 video typically results in 720p rectangular videos.

Single Lens Mode Video Resolutions

• Single Lens Mode offers a few video resolutions- 4k, 3.6k, 2.7k and 1080p.

• Recording in **4k or 3.6k is best when you want high resolution videos** for YouTube, or Vimeo for example. 3.6k is only slightly lower resolution that a standard 4k video on YouTube, so when you edit to 4k, the difference won't be noticeable.

But even if you record in 4k and want to post to Instagram, it's easy to lower to the resolution when you output your videos.

• Being able to record out of a single lens in 4k really steps the X3 up to the next level, giving you two cameras in one- both a 360 camera and a high resolution action camera.

2) FRAMES PER SECOND (FPS)

If you want to play your videos back in slow motion, selecting the right frame rate is how you make it possible. The number after the dash in 5.7k-30 (30 in this example) tells you how many frames are being recorded each second. Frames Per Second (FPS) refers to **the number of individual frames (similar to photos) the camera records each second.**

A video typically needs to play a minimum of 24 frames per second for the video to flow smoothly. Anything less than 24 frames per second and the eye can begin to distinguish individual frames (almost like a quick slideshow). When you record video at 24 frames per second and play it back at normal speed, this looks great. But when you record at a higher frame rate, say 60 frames per second on the X3, you can slow that video clip down to play 24 frames per second, which effectively makes that clip take 2.5x longer to play, giving it a slow motion effect.

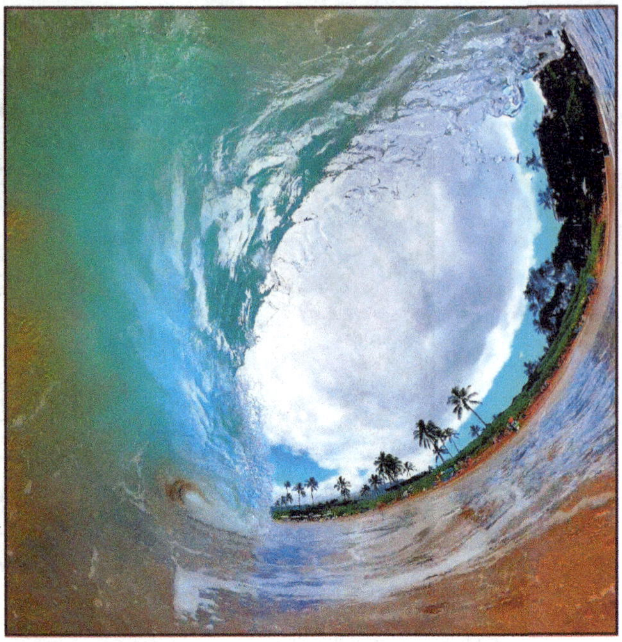

This fast action scene, for example, could benefit from slow motion.

Capturing footage at a higher frame rate (higher number FPS, such as 60FPS) is better if you may want to play back your footage in slow-motion, which a lot of action clips need to slow down the fast action. All of the video frame rates play back at regular speed by default, but with a higher frame rate, you have the option to edit it to play back in slow motion.

The important thing to remember is that as the frame rates increase, the available resolution is lower and advanced settings typically decrease, which is why 360 slow motion video is only available in 4k @ 60FPS. **Finding the right balance of resolution and frame rate for your scene is the secret to getting the footage you want.**

> **NOTE:** Many film and video makers display their footage at 24FPS because the on-screen "look" most closely matches film. Other professional video producers use 30FPS as a standard frame rate, arguing that 24FPS is only useful if you are transferring digital footage to film, which never really happens. 30 frames per second is better suited for viewing on televisions and computers. If you want to maximize the slow motion, use 24 frames per second, but we will use 30 frames per second for the video editing tutorials in this book.

The LowDown

- **360 VIDEO: Only select 4k @ 60FPS when you want slow motion.** The reduced resolution gives you less editing freedom but can be worth it when you want slow motion. 4k @ 60 FPS reframes to 1080p videos and maintains high quality. For Super Slow Motion, use 3k @ 100 FPS, but this resolution is only for 720p reframes.
- **SINGLE LENS VIDEO:** In Single Lens Mode, **3.6k @ 60FPS** is the best slow motion setting since 3.6k is only slightly smaller than 4k. You can reduce the frame rate later if desired.

> **TIP: ANTI-FLICKER- PAL (50Hz) vs. NTSC (60Hz):** Most countries outside of North America use a format for televisions called PAL instead of NTSC.
>
> If you are recording videos to upload to YouTube, you can use either NTSC or PAL. NTSC is a better choice because it offers a higher frame rate (60FPS vs. 50FPS) for slow motion. However, when filming under artificial lights (e.g., streetlights), the lights may flicker when using NTSC in a PAL country. *Auto, PAL (50Hz), or NTSC (60Hz) can be selected using the Shortcut Menu>Anti-Flicker setting.*

3) ASPECT RATIO

The third setting option affects the proportions of the video frame. The Aspect Ratio (16:9, 9:16, 1:1, 2.35:1, or 4:3) refers to the ratio (width: height) of the video image.

In Single Lens Mode, when you tap on the Main settings next to the icon, you can select between 16:9 Widescreen or 9:16 Vertical.

360 videos don't use an aspect ratio since they are recording a full sphere, not a rectangular video. When reframing 360 videos in the Insta360App or Insta360 Studio, the Aspect Ratio can easily be changed using the Aspect Ratio Icon.

There are **several ratios to choose from:**

16:9 Widescreen Aspect Ratio

• 16:9 Widescreen is a widescreen format for HDTV's, plasma widescreen TV's, and cinema screens. This is how you see most videos and movies. YouTube, Vimeo and Hulu support this ratio. Using this ratio gives your footage the cool widescreen cinematic look. The widescreen ratio also offers faster frame rates for smoother slow motion.

Widescreen is the best choice for most shots when you have enough distance between the camera and the subject to capture what you want in the frame.

9:16 Vertical Aspect Ratio

• 9:16 creates vertical videos, which fit the format of a lot of mobile content, such as YouTube Short, Tik Tok and Instagram Reels. If you are producing longer format YouTube videos, you don't want to use this format.

1:1 Square Aspect Ratio

• When editing your 360 videos in the Insta360 App or Insta360 Studio, this aspect ratio will conform your videos into a square format, which works well for Instagram posts.

2.35:1 Cinematic Aspect Ratio

• The cinematic aspect ratio creates a wider, shorter view for a movie-style look. On YouTube, this will show with a black bar at the top and bottom of the screen.

4:3 Standard Aspect Ratio

• The 4:3 Aspect Ratio is slightly taller than the 16:9 Aspect Ratio, which works well for point of view shots when you want to get as much of the subject in the frame as possible. YouTube, Vimeo and Hulu do support this ratio, but you will see black bars on the sides so the video fits in the widescreen player.

> **TIP:** If you plan to edit videos on a phone or tablet, the Insta360 App allows you to change the Aspect Ratio for 360 and Single Lens videos after you have recorded your videos. The available aspect ratios depend on the original media.

<u>The Lowdown</u>

• The versatility of recording in 360 means you can create a video in multiple aspect ratios, depending on where you plan to post it. This allows you to reframe multiple times without losing content from the scene, a bonus that in not possible with a traditional action camera.

• When editing single lens videos, adjusting the aspect ratio has less flexibility in terms of how much is cropped out since you don't have the full sphere to work with. Luckily, the X3 records more than what would be visible in a particular aspect ratio. For example, if you switch from 16:9 to 9:16, you will gain extra space in the top and bottom of the image.

TIP: Vertical Video. When recording Single Lens Videos, the Insta360 X3 offers the option to record video in either a Vertical or Horizontal Orientation. Although traditional video has always been recorded in a landscape format, vertical video formats have risen in popularity out of the simple fact that we typically hold our phones in a vertical position.

To give users the option of recording vertical videos in-camera, the Insta360 X3 can be set to record in either a vertical format or a horizontal format simply by selecting a 16:9 (Horizontal)Aspect Ratio or a 9:16 (Vertical) Aspect Ratio.

The vertical orientation also offers a unique opportunity to compose your videos far different than in the past. Some video shots can really benefit from the taller video frame for viewing on phones and tablets.

Of course, 360 videos can be edited into any format, both horizontal and vertical.

TIPS FOR CHOOSING RESOLUTION AND FPS

You can select the right settings for your shot (slow motion, regular speed, etc.) by choosing Resolution, and FPS.

When choosing the best combination of Resolution and Frames per Second, consider the following tips:

TIP #1: **Use the highest resolution** that suits your needs. It's always better to reduce a file size when you edit rather than increase it. In 360, 5.7k gives you the highest quality video files to work with, but if you need slow motion, use 4k @ 60 FPS. For super slow motion, you will have to step down to 3k.

TIP #2: When you edit your videos for slow motion during editing, **the audio speed will also be affected.** Just because you record at a faster frame rate doesn't mean you must use the clip for slow motion. You can play a clip recorded at any frame rate in regular motion video. If you play a video recorded at any frame rate to play back in regular speed at 24FPS, any extra frames will be removed without affecting the audio.

FOV FIELD OF VIEW (FOV)

The X3 offers two main field of view options- Default FOV and FOV+. These field of view options are available on the camera in Single Lens Recording.

Default FOV offers the highest resolution in Single Lens Mode, up to 4k @ 30 FPS. However, when you go to edit in the app, FlowState Stabilization and Horizon Lock are not available. This setting option is best for handheld or stationary shots without extreme movements.

FOV+ is only available in 2.7k resolution and lower, but **this option is for the widest field of view possible** in the Single Lens Mode with **FlowState Stabilization and Horizon Lock.** When you view the video on the camera, the footage will rotate with the camera. The magic of this mode happens when you bring it into the app. Using the menu in the three dots on the top right of the Insta360 App, you can apply FlowState Stabilization and Horizon Lock for some truly amazing video effects. **Use 2.7k @ 60 FPS** for the highest resolution in this setting with up to 2.5x slow motion.

FOV DIGITAL LENSES (FOV)

Changing digital lenses **affects the zoom, cropping and fisheye effect.** It's important to remember that they are merely "digital", obviously the Insta360 X3 does not have multiple physical lenses.

Digital lenses are definitely one of the most useful settings because you can easily swap lenses for a different look. After changing lenses, you can see a live view of how the different lens affects your shots. Once you understand the digital lens options, you can select the lens that best captures the perspective you want.

When using the **Default FOV in Single Lens Mode,** the Insta360 X3 offers four digital lenses to choose from- Action, UltraWide, Wide, and Linear. In **FOV+**, MaxView, UltraWide, Wide, and Linear are available. The lenses move from the widest perspective, (Action and MaxView) to the narrowest (Linear).

When **editing 360 videos** using the Insta360 App, four of the lens looks are available: UltraWide, Wide, Linear and Narrow. Since you are reframing the 360 video, these lenses are added after recording.

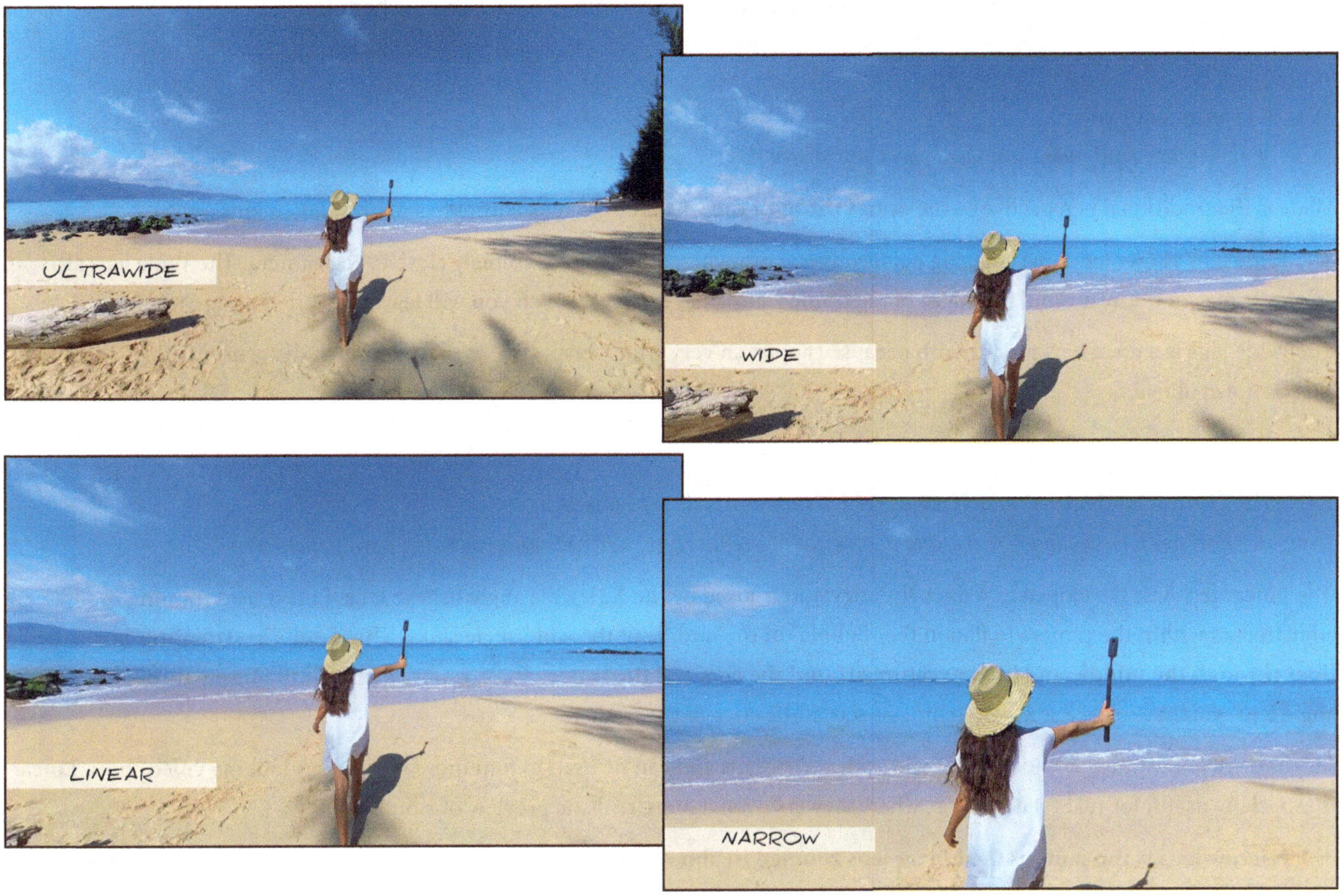

As you can see, changing lenses creates a different look, even when filmed from the exact same location. In the example, UltraWide is the most pulled back with extra curvature around the edges of the frame. Wide is your standard wide angle Insta360 lens. Linear creates a slightly more pulled back shot with straighter lines than Wide, while Narrow produces the most zoomed in shot with a reduced fisheye effect.

Let's look at a quick comparison of the digital lens options from widest to the most zoomed in:

MaxView (In Single Lens FOV+)

For POV Shots or Action Shots

MaxView gives you the widest angle possible in Single Lens mode. Along with Horizon Lock and FlowState Stabilization, this lens is **amazing for body-mounted action shots.** Because of the distortion caused by the extreme wide angle, you won't want to use this one for vlog shots too often. If you do, make sure to center yourself in the frame.

Action (In Single Lens Default FOV)

For POV Shots or Action Shots

Action is also a super wide-angle lens, which is **similar to MaxView**, but not quite as wide angle. Action does not have the Horizon Lock available, but it is compatible with Default FOV, so you have higher resolutions to work with.

ULTRA (ULTRAVIEW) (13mm equivalent)

For close up body-mounted shots, in tight spaces

Ultra is a super immersive wide angle perspective that captures as much of the surrounding scenery as possible. This is great for **up close or body-mounted shots** where you cannot get the entire subject in the frame. Max is also useful for dramatic landscape shots. Be aware that there is **distortion around the edges of the frame** when shooting in Ultra because it's such a wide angle shot. Ultra is **best for shots with the subject centered in the frame,** like paddling, biking or snowboarding.

WIDE (16mm equivalent)

Point of view shots, equipment-mounted shots, expansive skies

Wide is the traditional perspective of **the optics through the Insta360 X3 lens** without as many adjustments.

The Wide lens allows for some amazing perspectives, even when objects are extremely close to the camera. The Wide lens also creates some curvature around the edges of the frame (known as fisheye), which you will learn to work with in Step 4.

The Wide lens is a great lens to start with because captures a very wide perspective without as much distortion as the wider angle lenses shown above.

LINEAR (19mm equivalent) and NARROW (27mm equivalent) Lenses

Straighter horizon lines, good for real estate photography, architecture, portraits and recording family memories

The Linear lens was developed to reduce the curvature caused by the X3's wide angle fisheye lens. Linear uses framing adjustments within the camera to flatten the "bubble" in the middle of the wide angle shots. This results in **straighter horizon lines through the middle and less curvature around the edges of the frame.** Since your camera is still filming through a wide angle lens, you may notice some stretching effects at the edge of the frame.

Linear does not capture as wide of an angle as the Wide lens and can be used to film more traditional-looking video shots. Linear is especially useful for real estate videos, where viewers want a more realistic preview of a building.

The **Narrow lens is the same as the Linear lens zoomed in about 1/3.** Since you may notice a decrease in image quality when using the Narrow lens, the best recommendation is to film using the Linear lens and zoom in slightly when you edit if the video quality allows.

The drawbacks of the Linear and Narrow lenses are that you lose a bit of that action camera-style immersion we are used to seeing.

HORIZON LOCK (for 360 & FOV+ Single Lens Videos)

This is one of those features that will blow your mind at first. **Horizon Lock uses the full view from the lens you are recording with to keep the horizon level.** Even when your camera is rocking side to side or even rotating in a full circle, the horizon will stay level. When recording, keep the lens pointed in the direction of your subject. The Horizon won't stay level while you record, but once you bring it into the 360 App, you can lock the Horizon. It's amazing and offers so much potential for shots like you've never seen before!

For videos recorded in Single Lens Mode using the FOV+ setting, the Horizon Lock option is located in the Insta360 App using the three dots in the top right corner.

TIP: 360 Videos automatically have Horizon Lock applied. You can adjust the horizon using keyframes, like we will show you in Step 5- Editing.

PRERECORDING

PreRecording is a super convenient way to minimize the amount of wasted shots stored on your camera's microSD card by **recording the past.** When PreRecording is activated, you can set your camera to continuously record a small looping video of 15 or 30 seconds, but nothing is saved until you press the Shutter Button. After something happens that you want to keep, press the Shutter Button and the camera will then add the 15 or 30 seconds before you pressed the Shutter Button to the clip. The camera will continue recording until you press the Shutter Button again.

This is especially useful for short clips of trick shots, acrobatics, or fast action like wave cinematography.

Because the camera is continuously recording behind the scenes, you can expect it to heat up. In high resolutions, such as 5.7k, you may run into overheating issues, so use PreRecording wisely.

Another downside of using PreRecording is that since your camera is running continuously, battery life is the same as if you were continuously recording video. So, even though PreRecording is super useful for certain activities, it's not recommended when battery life is an issue.

PreRecording is available for Active HDR and standard Video Modes only and can be turned on in the Shortcut Menu.

FLOWSTATE VIDEO STABILIZATION

While the digital lenses affect the view, stabilization affects the motion. Insta360 cameras can easily be carried with you for an easy way to record life's great (and simple) moments. That handheld-style of filmmaking usually results in bumpy footage that's hard to watch. But not with FlowState's gimbal-like stabilization! When used with Horizon-Leveling, the Insta360 X3 quite possibly produces better results than a gimbal- pretty amazing!

The X3 offers two types of stabilization- in-camera and FlowState. In-camera stabilization is applied to all videos on the X3 (not on time lapses though), while FlowState is the highest quality stabilization.

How does it work? Both **in-camera and FlowState stabilization use the area outside of the frame to keep the video steady,** so as the camera rocks and rolls, the frame rotates like a gyroscope to stay steady. Since the X3 uses extremely wide-angle lenses, the camera has a lot of area outside of the frame to work with, resulting in super smooth videos. FlowState Stabilization can be applied to 360 videos and FOV+ videos because there is more area outside of the frame to work with.

FlowState is added automatically to 360 videos and Single Lens videos in FOV+, but can be turned off in the App using the top right menu (with the three dots).

TIPS FOR GETTING THE MOST OUT OF STABILIZATION:

The Insta360 X3 gives you gimbal-like stabilization without actually using a gimbal. Because your camera films stabilized shots but also needs the freedom to move-to pan, tilt, etc.- you can **give your camera's stabilization a helping hand by following these tips:**

- Stabilization needs an adequate amount of light to work properly. In daylight and in most outdoor scenes, stabilization works perfectly. However, **in low light situations,** such as in the evening or dark indoor environments, the level of stabilization applied will automatically be reduced. In low light, raise your maximum ISO to 800 depending on the lighting conditions (see more about ISO in Protune Video Settings which is coming up). Stabilization can create strange effects and ruin your low light shots, so if it's an important shot in minimal light, test it out first or turn it off.
- No gimbal or stabilization can totally compensate for erratic movements, so even though the stabilization will smooth out shaky shots, try to hold your camera steady and use smooth movements when possible.

ADVANCED VIDEO SETTINGS

These advanced video settings are available in 360 Video and Single Lens Video modes. The following Advanced Settings give you fine-tuned control over your video quality. These settings are mostly aimed towards experienced image-makers (which you will be soon) who want to affect the look and quality of their videos by adjusting manual settings. Now that you are learning how to maximize your camera, you should definitely use these settings.

• Advanced Settings in photo and video modes are slightly different. Advanced settings must be set individually for each mode, but these are important settings to instantly improve your video quality.

ADVANCED SETTINGS

You can select and change the following setting options within each Mode. The Advanced Settings are accessed by swiping left from the right side of the Touch Screen. In the Insta360 App, these settings are located to the right of the Shutter Button.

As you gain experience, you can customize these settings to your particular tastes and adjust them for changing lighting and action.

There are two groups of settings- **Auto and Manual.** In the beginning, use Auto, but as you see results and want control, use Manual when the scene requires it.

Start with the following recommended settings:

AUTO AND MANUAL SETTINGS

COLOR: Vivid Color

The X3 offers **three color profiles**- Vivid, Standard, or LOG. **Vivid Color** adjusts colors to add vibrance and give your footage pop. Insta360 really fine-tuned the color profiles in the Insta360 X3, making the raw videos ready to share without much color correction. To take advantage of these adjustments, choose Vivid Color. **LOG** produces videos with a more even tone if you prefer to adjust colors from scratch when you edit in post-production or if certain shots look over-saturated. Standard color creates processed videos, but the colors are more true-to-life in the Vivid color profile. If some of your shots filmed in Vivid Color are oversaturated (too much color), you can also reduce the saturation when you edit your clips.

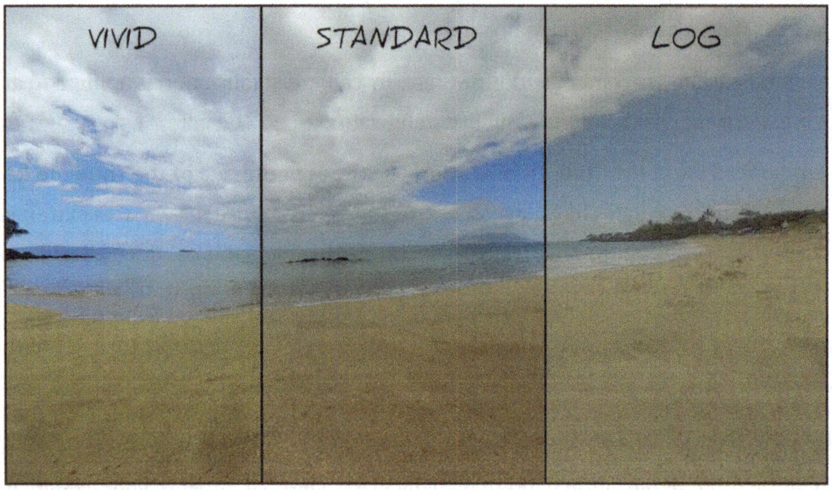

All videos shot in 360 Video on the Insta360 X3 at 5.7k @ 30 FPS. Notice how the colors are more vibrant when Vivid Color is used.

WHITE BALANCE: Auto

White balance refers to the color temperature- a lower number creates a colder (bluish) look and a higher number creates a warmer (more yellowish) tone. Unless you have experience setting white balance, or have a very specific need, Auto White Balance is your best option.

EXPOSURE VALUE COMPENSATION (EV COMP): 0

EV Comp tells your camera to make the footage brighter or darker than the exposure reading from the camera. You only need to adjust this manually if you are going for an overexposed or underexposed look, or if your camera is taking an inaccurate reading of a scene and you want to manually override the exposure reading. For example, if you are filming a scene with bright and dark areas, you can adjust the EV to +1 to brighten the scene one stop. Exposure Control (which you will learn about in this step) is an easier way to do this, but it's useful to be aware of how to use EV Comp.

Isolated Exposure: Off

Isolated Exposure tells your camera to read the exposure from each lens independently. This is useful when the exposure out of one lens is very different than the other lens. For example, if you have your X3 on the car dashboard, looking out the window will be very bright, while the view in the car will be darker. Turning Isolated Exposure on for a scene like this will result in a more balanced result.

Another option is to leave this off and turn your camera sideways so that each lens is viewing a similar set of mixed lighting conditions.

ADDITIONAL MANUAL SETTINGS

Shutter: Auto

The Shutter setting defines how long the camera's shutter stays open for each video frame. Auto is typically your go-to for Shutter settings. But, if you want to lock exposure to keep the Shutter open for a specific amount of time per frame, you can select a Shutter setting. A manual Shutter setting creates constant exposure throughout a video that will not adjust as lighting changes. Using Manual Shutter is not a good option in inconsistent lighting. In bright light, you will need to use a fast shutter with a low ISO to avoid overexposure. A faster shutter (1/240, 1/480, 1/960, etc) gives you sharper images with less blur. A slow shutter (1/30 or 1/60) can be used to create motion blur (a more cinematic look) but needs a stable mounting position to keep foreground objects in focus. You can see a live view of how the Shutter speed affects exposure by previewing the Live View on the Touch Screen.

Available shutter speeds, which vary based on the frame rate, are 1/30 sec, 1/60 sec, up to 1/8000. If you record at 60 FPS, the minimum shutter time available will be 1/60 sec. At 30 FPS, 1/30 will be available. These are the slowest shutter times possible to capture enough frames to match the selected frame rate. Not all Shutter speeds are available for every frame rate.

Because the available shutter speeds are typically slower than the Auto shutter speed, your shots may be overexposed in bright daylight.

ISO **ISO:** MAX 400 / MAX 800 for cloudy conditions

ISO is a very important setting that affects the quality of your videos. The ISO determines the sensor's sensitivity to light. If you want **to record the cleanest, noise-free video possible, tap into the lowest ISO available on your camera,** which is 100 ISO. Because a low ISO is less sensitive to light, it absorbs light more slowly resulting in high quality video without any noise.

As a general rule, if you want to use Manual settings, set the MAX to 400 and raise it to 800 for cloudy or shady conditions. Try to stay below 1600 ISO max, even for low light conditions. A high ISO (2000-3200) absorbs light more quickly and will allow you to record twilight or night video, but the video will be full of noise and appear to be low quality. Noise is all those little specks that make your footage look grainy like you may have noticed in videos you've recorded at night. Noise can be corrected to a certain extent when you edit, but it's best to start off with high quality, noise-free video.

3. HOW TO USE THE VIDEO SETTINGS

• Use **360 Video Mode when you want the most creative editing freedom.** Stick to 5.7k @ 24 or 30 frames per second for the highest quality footage. In Step 5, you will learn how to reframe your videos to a more traditional look or create unique 360 shots. You can also pull impressive still photos from the high resolution 360 video. For mobile edits when you want easy slow motion, step it down to 3k @ 60 frames per second.

• Even though Single Lens Modes do not record in 360, with the extra settings available, these modes are not to be overlooked.

• Choose **Single Lens Video Mode in FOV+** when you want to take advantage of **in-camera Horizon-Leveling and FlowState stabilization.** 2.7k @ 60 FPS is the best setting so you can play your footage back in 2.5x slow motion.

• For more **traditional action camera shots in 3.6k or 4k,** use the Default FOV in Single Lens Mode. Keep it at 3.6k @ 60 frames per second so you can slow it down 2x when you want to. Single Lens Video Mode's variety of lens options allow you to choose your settings for videos that are ready to share with less editing than spherical 360 videos.

TIP: Quick Button Preset Settings. The Quick Button on the side of the camera lets you quickly and easily change between your favorite setting combos. These can be customized to your tastes as you develop your favorite modes and settings.

To change the Preset Setting, select the Mode and Settings you want to use first. Then, press the Quick Button. If you want to overwrite one of the existing Preset names, use the Touch Screen to press the 3 dots next to the Preset you want to customize. Then follow the prompts to Overwrite the settings. Or, to add a new Preset name, tap the "+" symbol, select the preset name you want to use, and the current settings will be added to that preset.

TAKING PHOTOS

The Insta360 X3 offers a variety of photo modes to help you achieve amazing photos. There are a few of ways you can set up your camera to take photos, from taking a single 360 photo to taking a speedy burst of nine 360 photos in a split second. The 360 photo-taking capability of this camera is top-notch, and that is great motivation to understand the various shooting modes.

This section will help you understand when and how to use the photo modes when using your Insta360 X3 to take photos. Since each photo mode is used for a different purpose, we will cover each one in depth and provide the best settings and scenarios for that mode.

We will then look a little deeper into the more general photo setting options and explain how they can improve your photos.

Time Lapse Modes also offer a few settings that allow you to take photos as well, which we will cover in the next section.

PHOTO MODE

The Photo Modes are designed for you to operate your Insta360 X3 like a standard point and shoot camera, where pushing the Shutter Button tells your camera to take a photo. The X3 offers three different 360 photo modes, as well as a Single Lens photo mode. Let's explore each one a little deeper to give you the knowledge you need to choose the right mode for the shot you are trying to capture.

360 PHOTO MODES

The following three photo modes are available when you are in the 360 Modes- Photo, HDR and Burst. You will learn about **Interval** in the section on Time Lapse Modes, since this photo mode is used for photo time lapses:

PHOTO (360)- High Resolution 360 Spherical Photos

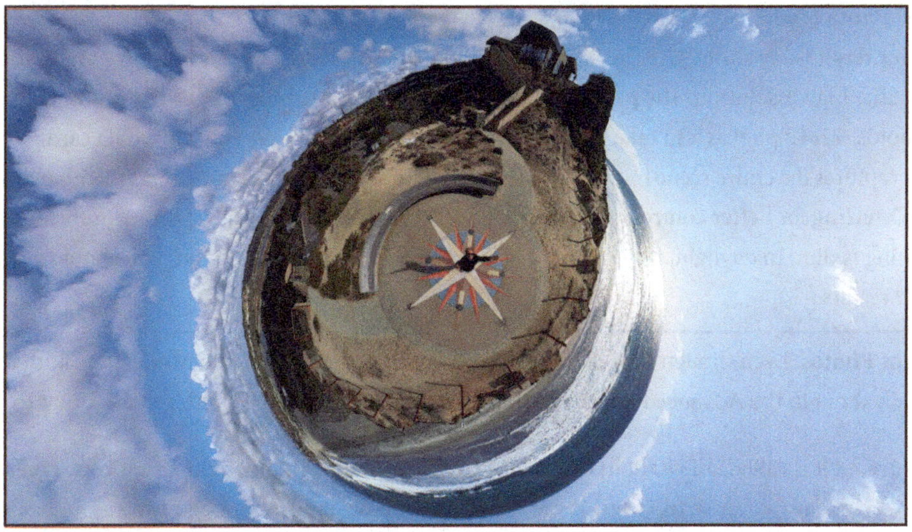

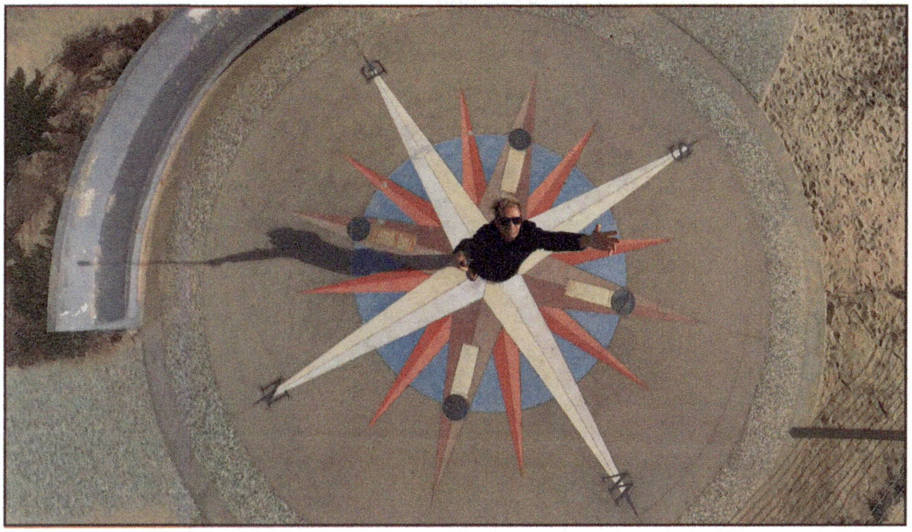

Two versions of a single photo taken with the Insta360 X3 camera using the 360 Photo Preset. The camera was held on the Invisible Selfie Stick fully extended. The first example shown is reframed into a "Tiny Planet" photo, but there is so much more you can do with your spherical 360 photos like you will see in Step 5.

360 Photo is the best 360 photo mode for **creating spherical 360 photos using both of the X3's lenses.** When you press the Shutter Button (or when you say, "Take A Photo"), your camera takes one photo. Spherical 360 photos can be shot at either a whopping 72MP (11968px x 5984px) or 18MP (5952px x 2976px). 72MP photos take longer to process due to the large file size, but these are excellent when you want the highest quality photos possible.

360 Photo Mode is such a fun, creative photo mode with tons of possibilities ranging from artistic to entrepreneurial. First of all, when mounted correctly (we will show you how of course), the mount disappears creating a magic floating camera effect. Secondly, the photos can be reframed for a wide array of looks from a single photo. You can extract a traditional looking photo or zoom all the way out for tiny planet photos. This is also the mode to use when creating panoramic photos for virtual tours.

Photos taken with the X3 **must first be opened with the Insta360 App or Insta360 Studio and exported** before being edited in another editing app. We will cover preediting and reframing styles in Step 5.

• When mounted on a pole, use the **self-timer** (set to 3 or 10 seconds) **or voice control** to reposition yourself further from the lens or to disappear out of the shot completely. We will teach you more tricks for composing your 360 photos in Step 4- Capture the Action.

• When taking photos in 360 Photo Mode, hold your camera steady and keep your subject around the middle of the frame for the least distortion.

Best Settings for 360 Photo Mode:

- Use **18MP photos for day-to-day shots.** Use the 72MP resolution when you are taking photos for an important photo shoot, where you really need the high resolution. The processing time is about 8-10 seconds for 18MP photos, compared to almost 20 seconds for 72MP photos. 72MP photos also take up about 4 times as much space on your microSD card.
- Because the camera captures the entire scene in the photo, **use the Timer** to press the shutter and then move away from the camera.
- **Adjust the Max ISO setting** for better control of image quality. Keep the ISO as low as possible for the highest quality photos but raise it if the lighting is dim. In daylight, if the camera is still and you are using the timer, you should always be able to use a low ISO and get good results.

TIP: Night Photos. Even though the X3 doesn't have a dedicated Night Photo Mode, the Shutter setting (which we will talk about in the Advanced Photo Settings) allows you to keep the shutter open for up to 120 seconds.

The **long exposure,** as it is called, allows light to **slowly "write" an image in low light situations.** Long exposure photos can be taken at dawn or dusk (up until about 30 minutes before sunrise or beginning around sunset) or during nighttime with your camera mounted in a stationary position.

When photographing a night photo, remember that **any incoming light that lands on your camera's sensor will leave its imprint.** A longer shutter time takes in more light and will create a brighter photo. Shorter shutter times let in less light and result in a darker photo. The right combo of ISO and Shutter will create magic night photos.

The best way to determine the exposure is to test out a few shots using a combination of low ISO and a long shutter until you achieve the desired exposure. For example, in a pure dark night sky, a 60 second shutter at 400 ISO will produce a proper exposure for star photos.

Use longer shutter times of 20 seconds+ to create stunning waterfall or ocean photos with a mysterious soft water appearance.

TIP: 360 Photos from video. Pulling still photos (called frame grabs) from your 360 videos is easy to do using the Insta360 App on your device. Pulling still images out of video taken in 360 Video mode results in high quality still frames which compare to the quality of the 18MP photos taken in 360 photo mode. Plus, in 360 Video Mode, you can record 30 frames per second for long periods and pull still photos from the video. Keep that in mind the next time you want photos and videos simultaneously.

HDR PHOTO (360)- Take High Dynamic Range 360 Spherical Photos

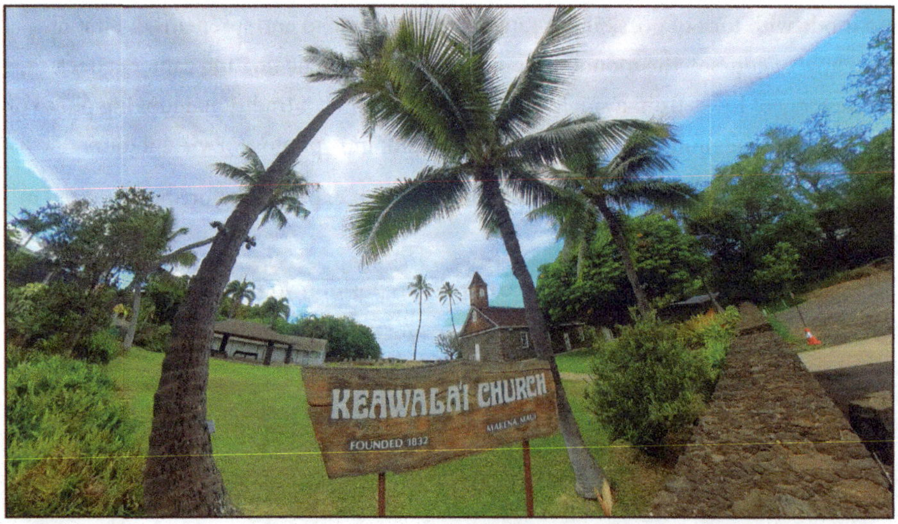

Photo taken with the Insta360 X3 camera using HDR Photo Mode with 7 frames. The camera was placed on the wall on the Invisible Selfie Stick with a monopod base by Insta360.

Use HDR photo **for high contrast scenes** when you want a better balance of bright and dark areas. This mode will automatically adjust the photo to balance out the contrasty areas. HDR Photo mode can create some **amazing scenic shots,** especially in imbalanced lighting such as at sunset.

• Since your camera will be taking multiple photos of the same scene, set your camera up on a tripod or in a stationary position. Use the self-timer set to 3 seconds (at least) to prevent camera shake from pressing the Shutter Button.

Best Settings for 360 HDR Photo Mode

• HDR Photos can only be taken using the **18MP resolution**. If you want to take advantage of the full 72MP photos possible on the X3, use Photo Mode (360).

• The **Auto Bracket setting** is unique to HDR Photo Mode and can be adjusted in the Advanced Settings (in the tray on the right). This setting tells the camera how many photos to take at different exposures, ranging from 3-9 photos. The photos will then be combined into a single HDR image, balancing out the tones from each image. For scenes with more contrast, select a higher number of images.

BURST (360) - Take 9 360 Spherical Photos In Less Than a Second

Burst Mode (360) is the mode to use **for high action moments** when you want to capture a split moment of time in a 360 photo. When you press the Shutter Button in this mode, the camera will automatically take a sequence of 9 photos over about half a second. As long as you can time that Shutter Button with the action (which can be kind of hard), you won't miss the best moment.

• Only 18MP photos are available in this Burst Photo Mode.

• To get the best shots out of Burst Photo Mode, use the self-timer set to 3, 5 or 10 seconds. Once you see the countdown end, make the action happen. Pick the best shot from the bunch and you'll have a winner.

• If you find it too difficult to jump into the shot at the right time, **an alternative to this mode is to record using 360 Video Mode.** You can then pull a still 360 photo from the video.

SINGLE LENS PHOTO MODE

When you switch over to the Single Lens Modes, only one of the modes is designed for taking photos using a single lens.

PHOTO (SINGLE LENS) - Easy-to-share traditional photos using one lens.

Single Lens Photo mode **takes a standard photo through just one of the lenses** with the press of the Shutter Button or by using voice control ("Take A Photo"). The photo resolution can be set to 36MP or 9 MP with a 16:9, 1:1, or 9:16 aspect ratio. SINGLE LENS Photo mode is **best used when you want to frame your photo for quick sharing** without having to reframe a 360 photo.

• The FOV (field of view) options are very useful for this photo modes to quickly change your composition between the UltraWide, Wide and Linear lenses. Max is an excellent lens for photographing mounted action shots.

OTHER PHOTO SETTINGS

Now that you understand the unique ways to use the Insta360 X3 to capture photos, these setting options can improve your photos even further.

The following Settings options can be changed in the settings dialog for each Mode.

SELF-TIMER

In photo modes, self-timer can be **set to 3, 5, 10, or 15 seconds** which makes it easy to **capture the perfect selfie or scenic shot without any shake**. Self-timer is also useful to give yourself a few seconds between pressing the Shutter Button and checking your composition before the camera actually takes the photo.

When taking 360 photos, the self-timer is really useful for moving your hand away from the camera or exiting out of the frame completely. The delay also allows you to stabilize the camera before taking the photo, preventing camera shake.

To use Self Timer, set the Self-Timer in the Primary Settings dialog. When you take a photo by pressing the Shutter Button, the self-timer begins counting down with the countdown time indicated on the Touch Screen.

ADVANCED PHOTO SETTINGS

Some of the Advanced settings for photos are the same as video settings, but there are a few differences.

The **following recommended settings** will help you understand how to use some of the advanced settings to improve your photos:

PHOTO FORMAT: INSP

There are three different photo formats available to choose from. 1) INSP is Insta360's standard compressed photo file, similar to a JPEG on most cameras. This is the easiest, smallest file to work with. 2) When INSP+RAW is selected, the camera will also create a RAW (.dng) photo file. A RAW file is uncompressed and allows a wider range of adjustments when you edit. (We will talk about how to edit RAW files in Step 5.) 3) PureShot is the third option. PureShot is Insta360's AI powered mode for creating HDR photos and vibrant night photos from a single image. Use PureShot when taking photos at Night or in low light scenes. PureShot photo files can only be edited on the Desktop App, but you can add PureShot effects in the mobile app.

WHITE BALANCE: Auto

As with video, Auto White Balance is your best option. White balance affects the color tone of your photos which can be fine-tuned using this setting option, but in general, the camera will produce an accurate reading of the correct white balance. For Night Photos, you may want to adjust the white balance to 2800 for a more realistic night tone.

EXPOSURE VALUE COMPENSATION: -.5

If you notice that your images are consistently turning out over or underexposed, use this to fine tune the exposure to your tastes. Most photographers typically like to avoid overexposing photos too much because the details in the highlights are easily lost, which is why -.5 is a good default setting.

ISOLATED EXPOSURE

This is the same as for video mode and can be used for extreme contrasts in lighting from one lens to the other.

MANUAL SETTINGS

SHUTTER: Auto

The Auto Shutter setting lets the camera select the right setting for your scene. However, by selecting a set ISO and Shutter speed, you can **manually set your exposure**. This is very useful for certain situations where you want a fixed exposure throughout a range of shots.

The Touch Screen shows a live view of how the manual Shutter setting will affect the photos, but it's still a good idea to take a few test shots to determine the correct exposure for your scene. A faster shutter setting (1/1000 sec. or 1/2000 sec.) lets in less light and will be most common for bright scenes.

The available Shutter settings are AUTO, 1/8000 second up to a long exposure of 120 seconds. This wide range of Shutter settings allows you to take photos anytime from the brightest daylight into the darkness. (Remember the tip in 360 Photo Mode about taking night photos because they are fun!)

ISO **ISO Maximum:** 100 in daylight/400 in cloudy conditions

As mentioned in the section on video ISO, when there is enough light to keep the ISO down, you will capture less noise in your photos, resulting in cleaner, clearer images. When you really want the highest quality photos and are photographing in bright light, set the max ISO to 100. In low light situations, setting a max ISO of 400 gives your camera a range so photos will also come out sharp. For Photos taken in really low light scenes, you may need to use 800 ISO, but at 1600 or 3200 ISO, the photos will have a lot of noise. Since you don't need to worry about stabilization, you can typically work with a lower ISO in photo modes.

That covers the powerful photo modes on the X3. Next, let's learn about how you can create stunning time lapses!

TIME LAPSE MODES

Time lapse modes give you the option to **set an interval between photos or frames,** spacing out your sequence over a longer period of time. The main purpose of time lapse modes is to create a gap between moments, which when played as a video, makes time speed up. With the ability to reframe your videos after you've recorded them, the Insta360 X3 is an **excellent tool for creating time lapses,** and there are several modes available to achieve the shot you want. And, as you will see, there are many ways you can use time lapse modes to create truly unique time lapse videos!

Time lapses make an artsy addition to a video. Learning the art of creating a visually-appealing time lapse takes some experimentation to get right, so don't be discouraged if you don't get a great one on your first try. Since time lapses are a new style of filming for many people, we will first cover some general time lapse tips and then move into the different time lapse modes (along with recommended Setting Presets) for recording your time lapse.

The following **general time lapse tips will help you with some of the more technical aspects of recording a memorable time lapse:**

- For a still time lapse, use a tripod or **set your camera in a stationary position so the scene creates the movement instead of your camera.** Any movement of your camera will be intensified.
- TimeShift Video mode (as you will learn about next) makes it **easy to film a stabilized hyperlapse, where your camera changes position throughout the clip.** There are lots of creative ways to film captivating hyperlapses as you will learn about in the TimeShift Video section.

360 TIME LAPSE MODES

The Insta360 X3 offers three time lapse capture presets to choose from depending on what you are filming and how you want to capture it. All three of the following options are available only in 360 recording. There are no time lapse options in the Single Lens Modes.

TIMESHIFT VIDEO- Stabilized Time Lapses for Action Shots and Hyperlapses

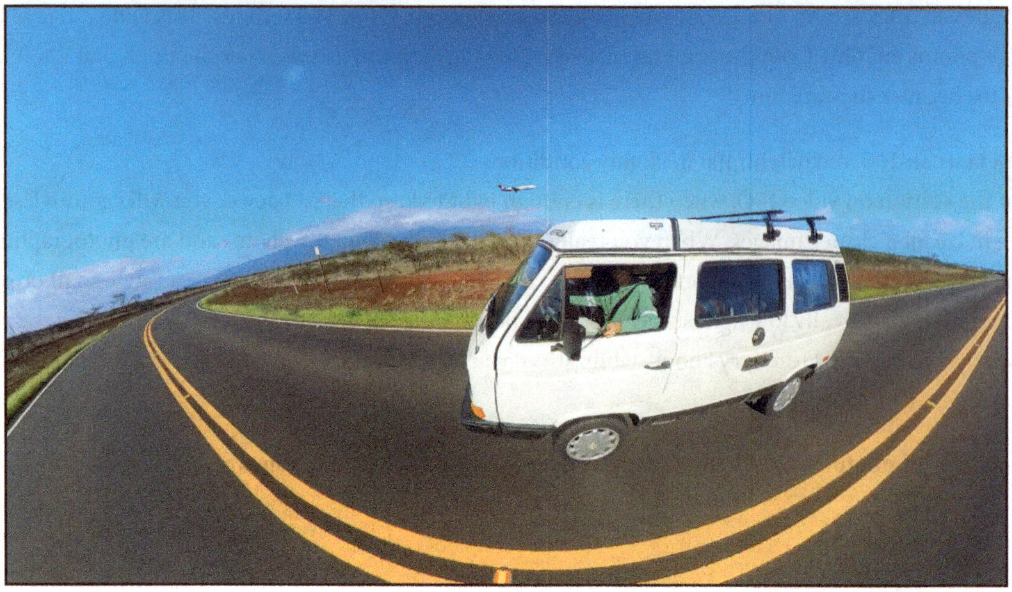

Video shot on the Insta360 X3 in TimeShift Video (360) Mode. The camera was mounted on the Invisible Selfie Stick and handheld. Reframing was used to rotate the view as the TimeShift played. You will learn how to reframe in Step 5-Creation.

TimeShift Video combines **time lapse video with stabilization** to produce smooth time lapse videos in-camera. TimeShift Videos record in 360 Mode for full spherical hyperlapse videos with infinite possibilities for creative edits.

TimeShift Mode uses shorter intervals between each frame than the standard Time Lapse mode, making this **the go-to mode for action time lapses and hyperlapses.** With the shorter intervals, **TimeShift Videos require less recording time** to create stunning videos and the stabilization smooths out movements so it's almost fool proof.

However, because TimeShift videos are typically recorded in much shorter time periods than a scenic time lapse, you can **recreate this effect easily from a 360 video** using the Insta360 App or Insta360 Studio. So, because of that, you're better off recording videos and creating the TimeShift video when you edit.

Whether you create TimeShift videos **using this mode or plan to edit your video into a TimeShift video,** there are some techniques you can apply while filming to make these Time Warp videos even more eye-catching!

These tips will help you create the best TimeShift Videos possible with the X3:

- Use a Max ISO of 400 for sharp shots in a variety of lighting conditions.

- In general, set the WB to Auto, but if you are filming primarily during daylight you might want to set it to 5500K to keep it consistent throughout the TimeShift.

- Use TimeShift Mode to **create easy, smooth hyperlapses.** A time lapse is typically filmed from one stationary position or with very slow, steady movement. A hyperlapse, on the other hand, is filmed by moving the camera's location between every frame. To film a hyperlapse, keep your camera fixed at a certain point or object as you move. The short intervals between frames in TimeShift Video Mode allow you to move relatively quickly and record a lot of shots in a short period of time. Since TimeShift videos are recorded in 360, it's easy to edit a video to bring out the focal point or subject.

- TimeShift Video is especially useful for activities that may be a little slow to watch the full duration, such as a long hike, a scenic bike ride, a road trip, or a long distance paddle. When you know you want to play your video back faster than real time, use TimeShift video to record your journey.

- TimeShift videos **can also be filmed at night** but you will need to use a higher ISO of 400 or 800. For super dark scenes, a high ISO of 3200 or 6400 is required, but the videos will have a lot of noise. The X3 also has the StarLapse option, which offers a long exposure and works better for night shots when the camera is stationary.

- Bring TimeShift videos to life by **using the filming techniques** you will learn in Step 4- Capture Your Action, such as tilting and panning. Just remember to draw out your movements much, much longer than normal since your movements will be played back faster than real time. Since you are recording a full 360 sphere, some of these movements, such as panning and tilting, can be added when you edit.

• TimeShift Video Mode is **available in 5.7k resolution.**

A Few TimeShift Video Ideas

• **Stand out in a crowd.** Film a crowded area with the subject standing still in the middle. When you speed it up, the subject will look like an island of calm in a sea of movement.

• For travel clips, film a hyperlapse as you **circle around your favorite landmark** or scene focusing on a central object to give your viewers a quick, stabilized 360 degree view for the full experience.

• Follow behind or look back at a subject while you **walk through a crowded scene.** Your subject will be the only thing that seems to stay steady.

• Film your journey **down a path or trail** for a quick lead up to your action clips.

TIME LAPSE MODE- Automatic Time Lapse Videos for Longer Duration Activities

In Time Lapse Video Mode, your camera records one frame (similar to one photo) at the selected interval (.5 sec, 1 sec, etc.). The individual frames are then **automatically combined** when you view this video in the app and **played back as a video** at 30, 25, or 24 frames per second. Because the frames are spaced apart and then played back quickly, there are gaps in between moments and time appears to speed up, creating the time lapse effect.

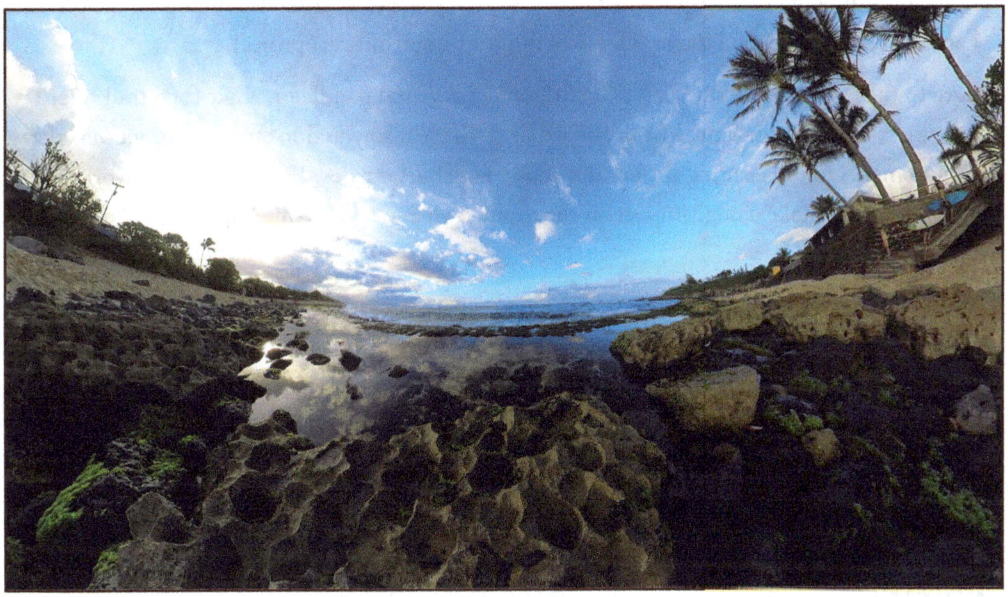

Shot on the Insta360 X3 using the Time Lapse 360 Video Preset with a 1 second interval. The camera was mounted on the Invisible Selfie Stick as a tripod for a stationary scene. A low ISO of 100 was used in the low light scene for the highest quality since the camera was stationary.

Time Lapse Video Mode is **best used for stationary time lapses when you want the highest resolution possible.** The X3 can record these time lapses in full 8k resolution, which can be viewed as a VR video or easily be reframed into a high quality 4k reframed video.

This section will help you with some of the more technical aspects of Time Lapse Video Mode.

• In Time Lapse Mode, the interval setting defines the amount of time between each frame of video (or photograph depending on your settings).

• Unlike standard video mode, the time counter **displays both the recording time (on the left) and the playback duration (on the right)**. Depending on the frame rate you select, **to capture one second of video, you need to record 30/25/24 frames.** The playback time counter will remain at 00:01 (1 second) until you have recorded enough frames for 1 second of video.

The following chart tells you how long you need to record at each interval for 1 second of video at 30 FPS:

INTERVAL (in seconds)	RECORDING TIME REQUIRED FOR A 1 SEC. CLIP
2	1 minute
5	2.5 minutes
10	5 minutes
30	15 minutes
60	30 minutes

• Choose a **short interval** (.2 to 5 seconds, but 2 seconds is a good go-to) for a scene with **continuously moving action** like waves lapping on the beach or traffic in a city for example. A short interval is also useful for **an event that happens over a relatively short period of time,** like a sunrise or preparing your gear to go ride.

• Choose a **long interval** (10, 30 or 60 seconds) for a scene where there is **not a lot of movement**, like slow-moving clouds, or for **longer duration events**, like road trips, construction or long art projects.

• When recording a Time Lapse video, **mount your camera in a stable position and add slow movement when you edit** for a professional look, which is easy to do with spherical 360 footage.

• Since the battery will run out after a couple hours max, you will need extra batteries or an **external power supply for longer interval time lapses.** When recording extended time lapses, you can use a 5V/3A external USB power bank to power your camera through a USB cable while you record. If using an external power supply with your camera mounted, open or remove the side door for access to the USB port on your camera.

• The **ISO can be kept low** since there will be no (or little) camera movement and this preset is for use in daylight time lapses. Raise the ISO to 800 for low light scenes, such as sunsets or sunrises.

• 360 Time Lapse Video Mode can be filmed in **8k or 5.7k resolution at 30, 25, or 24 FPS.**

A Few Time Lapse Ideas

• Record **an artist making a drawing or a painting.** Choose a relatively short interval depending on the predicted time of the project. Use Time Lapse Video at a 2 second interval for a 45 min to 1 hour-long art project.

• **Mount your camera** to the back of a standup paddleboard for a downwind paddle or a bike for a trail ride. Try to set up your shot with a stationary object in the foreground, like the board or a bike seat so you don't make your viewers dizzy. The moving action in the background makes for an exciting action time lapse. For shorter rides, TimeShift Video works well for this type of shot since the stabilization will make a huge impact on the result. Switch over to Time Lapse Video mode at 2 sec. or 5 sec. intervals for long duration rides.

• **Film the sunset at the beach or lake.** Try to choose a day with scattered clouds, and it's even better if the clouds are moving quickly through the sky. The movement of the clouds, the reflections on the water and the dwindling light will create dramatic time lapse clips. Try to frame your shot so there is some movement close to the foreground, like breaking waves or trees blowing in the wind. Because of the wide-angle lens on the Insta360 X3, movement that is far off in the distance won't make much of an impact.

INTERVAL- Takes 360 Spherical Photos at a Set Interval

Interval Mode allows you to select **a set interval between photos,** anywhere from 3 seconds up to 120 seconds. Once you press the Shutter Button, the camera will continue taking photos until you press the Shutter Button again. This is essentially a photo alternative to Time Lapse Mode since the end output is a photo file, rather than a video. Choose between 72MP and 18MP photos for the ouput.

When using Interval Mode, instead of Time Lapse Video, the end result will be a batch of individual photos. Some people prefer to batch edit photos for more control, combining them into a time lapse video after the photos are tuned up.

Interval Mode will not likely be your most commonly used mode, but there are a few creative uses for this mode:

• **Shoot photos of yourself in action** by setting the interval and pushing the Shutter button before you want to start taking pictures. You don't have to worry about pressing the Shutter Button at the right time and you can continuously take photos of yourself or a group. The individual photos will be available for you to choose from. This technique results in a lot of photos, but you can select the best of the bunch and delete the extras. This is a **great option for 360 photos** since you have more time to move further from the camera while continuously taking photos.

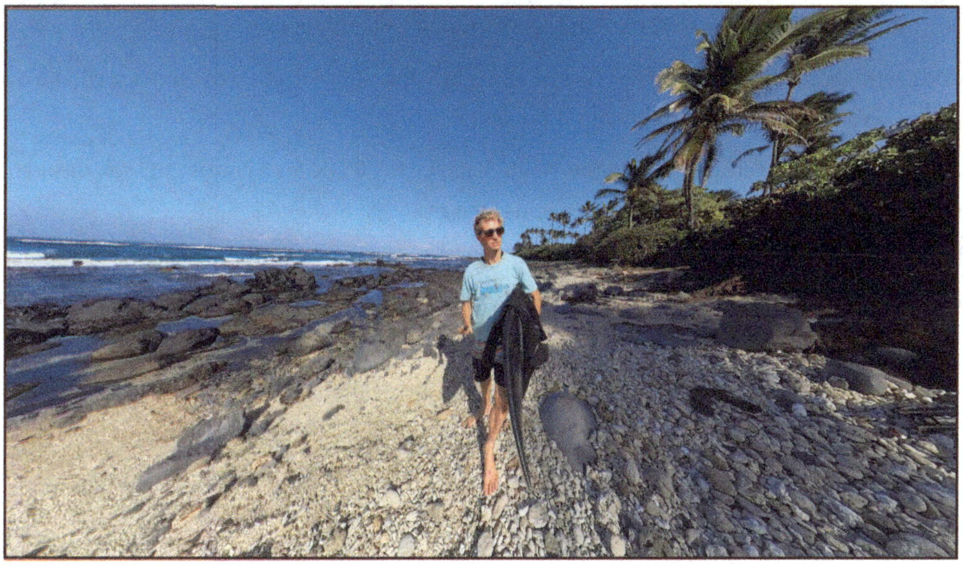

Photo taken with the Insta360 X3 camera using the 360 Interval Photo Mode to get a batch of high res 72MP photos to choose from.

- When using Interval Mode to take photos, **use a short interval (such as .5 seconds)** so you don't miss the best moments. The INSP photos taken in this mode don't require much space on your camera's memory card.
- If your end goal is **to create a time lapse video, it's typically better use Time Lapse Video Mode.** Merging 360 photos into a timelapse with the ability to reframe the timelapse is more complicated than reframing a timelapse created using the video format.

STAR LAPSE MODE- Night Lapse Photos taken at an Interval to compile in the mobile App.

Star Lapse Mode gives you the settings you need to record **long exposure night lapses.** Whether you want to create a night time lapse with trails behind the stars or a city buzzing at night, Star Lapse Mode is the mode to use. When using the Star Lapse Mode, your camera will take a batch of long exposure photos (long exposure means that the camera's "eye" stays open longer than normal, absorbing more light onto the sensor). These photos will then be automatically combined in the Insta360 App to play back as a video (when used in conjunction with the Star Trails Shot Lab). You can also access the individual photo files if you transfer them directly from the microSD card.

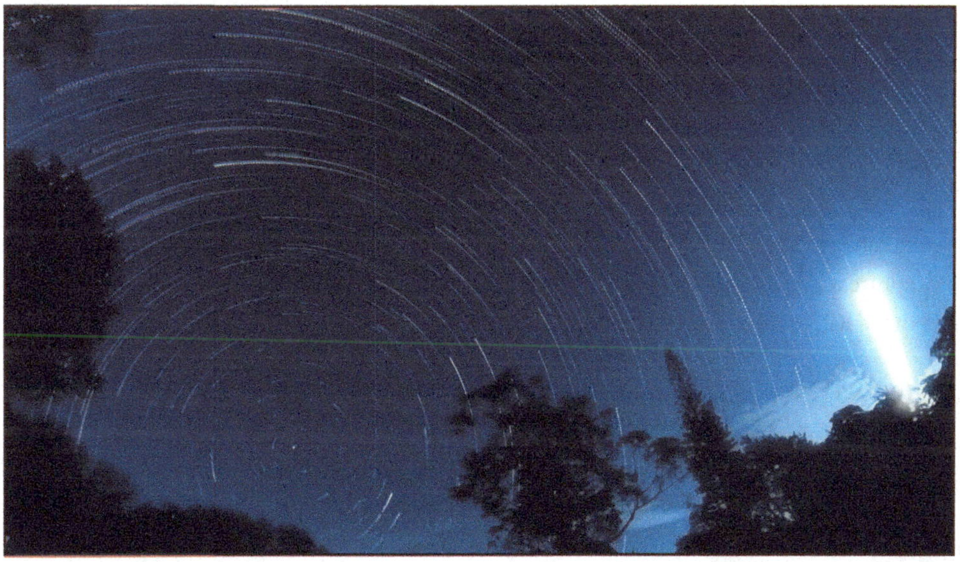

Star Trail video taken with the Insta360 X3 camera using the Star Trail Mode. The camera was mounted on a monopod and connected to a power bank.

These tips will help you create the best nighttime Star Lapse Videos possible with the X3:

- Mount your camera on a tripod to **keep the camera steady** for the night. **Connect to external power using a USB cable.** Since the camera side door will be partially open, only film on nights without a chance of rain.
- The Star Lapse Mode settings are **designed only to record at night.** Once dawn arrives, the footage will be overexposed. Star trails take a long time to create so record for a full night if possible.
- For **the easiest setup,** use the Auto settings, but set the White Balance to 3000K for a consistent look throughout the night. If your footage looks grainy because of the auto ISO, however, use the next tip.
- If you want **the highest quality possible**, use the Manual settings to get as low of an ISO as possible. For a dark night sky without a moon, 400 ISO with a 60 second Shutter Time, and White Balance set to 3000K is a good starting point. You will need to test the exact settings for your scene though.
- A **clear night is ideal** for Star Lapse Mode. If clouds roll in during the night, they won't necessarily ruin the shot. Clouds can create a nice visual effect as they come and go through the night, but a totally clear night is best.
- After recording a star trail shot, the clip has to be **edited in the Insta360 App using the Star Trails Shot Lab.** The export size is **limited to 1080p** in the App. We will talk about using Shot Labs in Step 5-Editing.

LIVE STREAM MODES

The last of the modes are the Live Stream Modes which are only accessible when your X3 is connected to the Insta360 App on a phone or tablet. After connecting to the Insta360 App, choose your streaming service. (Live streaming is available to Facebook, YouTube, Kwai, or through RTMP.) The 360 live stream will be automatically stitched in the app and streamed to your audience.

Two modes are available, Live and Live Reframe. Both modes are used to live stream directly from your camera through the app. The difference between these modes is that **Live offers your viewers a full 360 view that they can pan around,** while in **Reframe Live, you set the view and can change it**. The viewers can only see the view that you choose.

> **TIP: Android Users.** If you are using an Android device, the best option is to connect the X3 via cable using the official Insta360 Transfer Cable for Android. Utilize the camera's WiFi signal for streaming the video.

SHORTCUT MENU and CAMERA SETTINGS

The Shortcut Menu allows you to change various settings on your camera with the tap of a button, as well as go deeper and change some of the general camera settings.

After powering on your camera, Swipe Down on the Touch Screen to bring up the Shortcut Menu. Some of these options, such as Voice Control, QuikCapture, Connections, and checking your camera's firmware version were discussed in Step 1. Some of these are self-explanatory, but we will cover the ones that could use explaining.

SHORTCUT MENU ICONS

VOLUME CONTROL

Tap this turn on/off the camera beeps.

INDICATOR LIGHT

Tap this turn on/off the indicator lights on the front and back of the camera. This is useful for night recording to prevent the light from flashing on your subject.

Vibration

Tap this turn on/off the vibration which indicates when you begin recording.

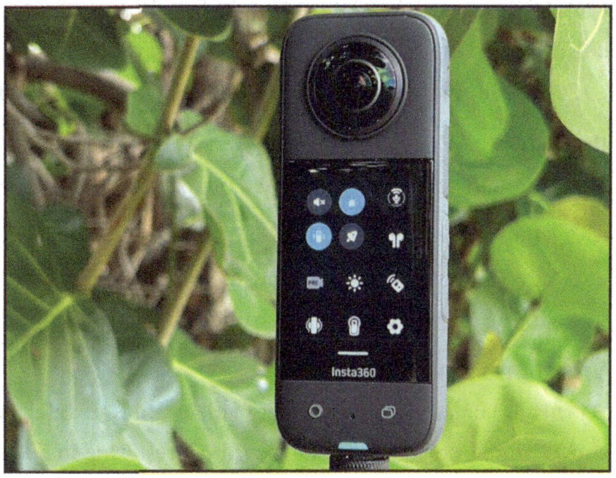

The Shortcut Menu

QuickCapture

Tap this turn on/off QuickCapture

Audio Setting

Select between Direction Focus, Wind Noise Reduction and Stereo.

Direction Focus is the perfect setting for reframing your 360 videos. When you reframe 360 videos in the Insta360 App or Insta360 Studio, the audio will be enhanced according to the camera's view. In Single Lens Mode, the audio will be enhanced based on the lens you are using.

The **Wind Noise Reduction** setting is used to boost your voice and reduce background wind noise. This can be useful when you want to record usable audio directly on the camera in windy conditions. Also, if you want to record nature sounds (such as the ocean or a waterfall), set this to Off to prevent the camera from trying to correct the natural sounds.

Stereo audio means the microphones record multidirectional audio. Take advantage of the four microphones on the X3 to record audio in full 360 stereo audio. Stereo audio is also available for Single Lens Video Mode.

We will talk more about audio options in the Audio section in Step 4.

AirPods

Tap to connect to AirPods, which can be used for recording audio externally. This is a simple and easy way to record external audio using Bluetooth.

Pre-Recording

As we talked about before in the video settings, use this to record a 15 or 30 second clip before pressing the Shutter. Once you press the Shutter, the preceding time period will be saved and added to the current clip. Since this continuously records, only use this setting when battery life is not a concern.

Brightness

Adjust the Touch Screen brightness.

Bluetooth Remote

Tap to connect to the GPS Remote Control.

Lens Guards

Tap this to apply Stabilization and adjust the stitching when using Lens Guards. This setting can also be applied in the Insta360 App or Insta360 Studio.

DiveCase Mode

We talked about using Dive Case Mode in Step 1. Once you unlock the Touch Screen, Dive Case Mode will be disabled.

Settings

Tap this to enter the Settings Menu. Let's look at a few more of those items now.

The following options are found in the Settings Menu:

Camera Info

This displays your camera name, serial number and firmware version.

USB Mode

Use this to change the USB-Mode if you are using the Quick Reade or connecting your camera to an Android device. The default U-Disk Mode allows your camera to be read by a computer when connected.

Bluetooth Wakeup

This allows the camera to be turned on using the remote or the App, but it increases battery consumption.

Anti-Flicker

Use Auto (to have all settings available) or select 50Hz for PAL/60Hz for NTSC. NTSC will give you faster frame rates, but will cause flickering when used with artificial lighting in countries that support PAL.

Video Bitrate

Set this to High for the best quality videos. Bit Rate refers to the amount of data being processed per second and the High setting allows more information to be recording, resulting in better quality video.

Video Sharpness

Sharpness refers to the amount of digital sharpness that is added to your footage in the camera after it is shot. Sharpness adds more defined edges to the images. Videos with too much sharpness applied don't look good, especially when viewed on a larger screen. Choose between Highest, High, Medium, or Low. Low is best if you want to sharpen videos when you edit. Medium is the best option if you don't want to manually sharpen the videos.

External Mic Gain

Raise the audio levels if you are using an external mic.

Auto Sleep

Set the camera's Touch Screen to sleep (turn off) after a set period of idle time.

Auto Power Off

Set the camera to completely turn off after a set period of idle time.

Language

Set the camera's language (for screens and settings, etc.)

SD Card

This shows the amount of SD card memory used out of the total capacity. You can also Format the card here, which will delete all files.

Customize Button

Change the function of the Lens Button to either switch lenses or work as an additional Shutter Button.

Gyro Calibration

Use this to reset the internal gyroscope. This ensures accurate stabilization for your camera. If you notice that the horizon is not level in your shots, set the camera upright on a level surface and run the calibration.

Factory Reset

Reset everything to the factory settings.

Congratulations, you've made it through the technical settings step. You are now ready to move on to Step 3 where you will learn how to mount your camera!

STEP THREE

MOUNTING

Set Up Your Camera To Work The Angles

The quality that really makes Insta360 cameras stand out is their ability to be mounted in unique locations to capture angles that used to be unwieldy or nearly impossible. The Insta360 X3 adds even more magic to the mounting mystery by making the mount disappear when mounted properly. We will show you everything you need to know about mounting the Insta360 X3.

When it comes to mounting your X3, think outside the box. Look for angles you've never seen before. Or use others as inspiration to record your life in a way you never thought to do before. **Getting creative with mounts is what will make your footage truly unique.**

There are hundreds of ways to mount your Insta360 X3 and the possibilities keep expanding as creative users come up with new mounts and new mounting techniques. If there is an angle you can imagine, there is a way to capture it with your Insta360 X3. And the best part is that this camera is so small and lightweight that it's hardly even noticeable as you carry it along with you on your big (and small) adventures.

So, study up and learn the best ways to mount the Insta360 X3 so you can decide how to best capture your point of view. The mounting examples also provide more insight into choosing your modes and settings for particular shots.

This chapter begins with the basic elements- **the camera and the 1/4" mounting base** - and then gets **deeper into the wide variety of mounting techniques you can use with the Insta360 X3** to get the angles you want to capture.

We will **primarily focus on mounting for 360 videos and photos** since 360 video offers truly unique mounting options. For Single Lens Modes, the X3 can also be mounted to any other mount just like you would a traditional action camera.

Other companies besides Insta360 make versions of many of these mounts, however this book only points out other available options when they offer something unique from the original Insta360 mounts.

GETTING STARTED

The Insta360 X3 is **part of a unique mounting system** that makes it stand apart from other traditional cameras and is part of the reason you are able to create truly mysterious shots. The base of the X3 has a **1/4" tripod mount** for attaching your camera to a variety of mounts. This basic combination is the starting point for a whole world of fun, creative mounting techniques. The Insta360 X3 takes it to the next level of imagination with the ability to make the mount disappear!

Let's get started with one of the most exciting and creative aspects of using your Insta360 X3- mounting!

MOUNTING BASICS

First, let's cover a few of the basics you should know before mounting your Insta360 X3.

USING THE INSTA360 X3 IN THE WATER

Yes, the camera is waterproof! But underwater videos are not clear without the Dive Case.

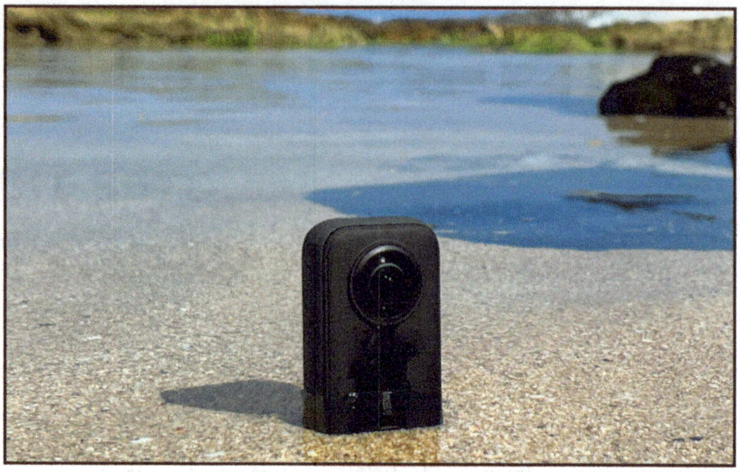

- The Insta360 X3 is **waterproof to 33' (10m) without an additional case as long as the side door and battery compartment is closed.**
- However, to create the 360 spherical image, Insta360 had to use two curved lenses. Unfortunately, **the curved lenses** required to create spherical photos and videos **don't produce clear imagery underwater.** In both 360 Mode and Single Lens Mode, photos and videos recorded underwater will turn out blurry.
- To record in 360 Mode or in Single Lens Mode underwater, **use the Insta360 X3 Dive Case down to a depth of up to 164' (50m).** The Dive Case is designed with the 360 stitch lines in mind. Although you will see the base of the case in your 360 images, the rest of the 360 image will be stitched perfectly and in focus. As you film, try to keep yourself or your subject away from the base of the camera. Also, when you go into the Insta360 App to edit, in the additional settings in the top right corner, make sure to **select "Stitch for Dive Case"** in the options. There are two settings, one for underwater and one for above water since water affects the way the footage needs to be stitched. The Dive Case has a hydrophobic lens coating to repel water drops, so follow the suggestions in the next tip for the Dive Case as well.

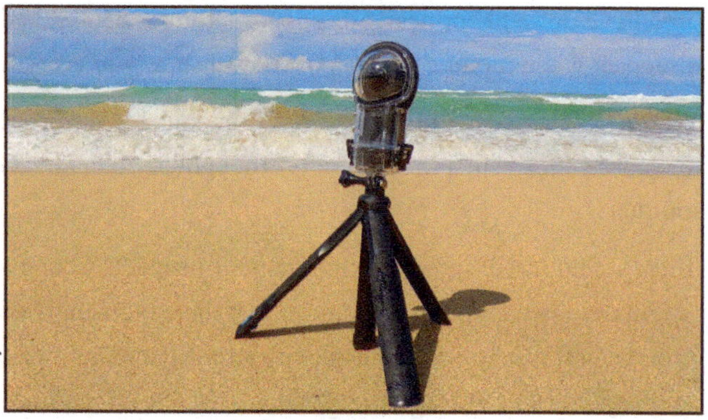

The X3 Dive Case

- **If you are in the water recording above the surface,** water drops on the lenses quickly ruin a shot. Water drops on the lens are visible in your shots and any water around the edges of the lens will affect the stitch line in 360 spherical photos and videos.

The first step is to make sure the lenses are clean and buffed out. **To prevent water drops from collecting on the front of the lenses when your camera gets wet, buff the lenses with a clean, soft cloth before going in the water.** Any oil or grease on the lenses will cause water to collect.

The second step is the try to get the water drops off. You want to **try to keep the lenses out of the water as much as possible to reduce water drops**. If the lenses get wet, **shake your camera or blow on it lightly to try to remove the water.** With the X3, you want the water drops to fall off or repel off the lenses as quickly as possible. The hydrophobic lens coating helps to repel water, but with the extra care, you will get nice, water drop free shots.

Unfortunately, licking the lens like you can do with some standard action cameras doesn't work well for the X3. The layer of water remaining on the lens, especially around the edges messes up the stitch lines.

- The Insta360 X3 **does NOT float. Whenever you go in the water, use a floating handle** to float your camera. The SP Gadgets floating handle has a ¼"-20 top screw attachment to add an extension pole or mount directly to your X3. Insta360 also makes the Floating Hand Grip with a GoPro-style mount on top, but this handle is not extendable. You don't want to watch your camera sink out of your hands and into the depths! Also, be sure to **lock the Touch Screen by pressing Dive Case Mode** on the Shortcut Menu to prevent water from changing the settings.

PROTECTING THE LENSES

- The two lenses can scratch easily, so they need extra caution. When storing your camera, make sure to keep it in the included neoprene case. Any scratches on the lens can potentially be seen in the shots. For extra protection, you can add the **Sticky Lens**

Guards. These protect the lenses, and if the Lens Guard get scratched, you can easily replace it. If you decide to use the Sticky Lens Guards, make sure to change the Stitching setting in the Shortcut Menu for the Sticky Lens Guards.

THE SIDE DOOR

• The USB **side door is removable.** When the side door is open or removed, you can **access the USB port** to charge the battery, transfer files or connect to external USB power while your camera is mounted. However, **your camera is NOT waterproof with the door removed.** Also, be careful removing the side door because it breaks easily.

• **As long as the battery compartment and USB side doors are securely closed,** the Insta360 X3 camera itself is always ready for filming in the water above the surface (because it won't be in focus underwater without the Dive Case).

• Whenever you finish accessing the side door or battery compartment and are ready to start using your camera again, **make sure both doors are closed completely** with the tabs closed so you don't see any orange. You'll notice a small rubber gasket around the openings of these doors. That gasket seals water out of your camera. Keeping the doors securely closed is essential to keeping your camera waterproof.

• When your camera is new, the door should close easily and securely latch. But after time, some dust or dirt may accumulate making it a little rougher to close. Make sure to keep any sand or dirt off the gasket by cleaning it periodically.

DIVE CASE FOGGING ISSUES

• When using the **Dive Case,** fogging (moisture on the inside of the case) is likely to occur, especially in **humid environments.** First of all, use the small foam inserts that come with Dive Case. These are designed to absorb extra moisture. If you notice fogging on the inside of the housing, stop recording when it is convenient to let the camera cool down. As the camera cools off, you will notice that the moisture inside of the housing will start to dry out. The fogging is **caused by the difference in temperature** inside of the glass and outside, just like a car window fogs up. The most effective way to minimize fogging is to use the provided anti-fog inserts. These small foam inserts absorb some of the moisture and can only be used a few times. You can also minimize the condensation by letting your camera adjust to the climate before recording. For example, if you come out of an air-conditioned room into 100-degree heat, wait a bit for the air inside the dive case to adjust. If you are going into the ocean or lake, put the camera underwater for a minute to let the temperature adjust.

• If you are experiencing fogging problems during a filming session, try to record only when you need to. The camera heats up when actively recording. **Use QuickCapture** to reduce standby time, which will allow your camera to cool off and give the moisture time to dissipate in between shots.

A FEW MORE MOUNTING TIPS

• **Keep it simple.** You could get overambitious and buy every mount (because yes, there is a fun, creative use for almost every mount), but simplicity is going to be your best friend. When you are out having fun, whether it's biking, snowboarding, or just cruising around, the key is to be in the moment. Prepare beforehand. Select two or three mounts to take with you to avoid spending your time fumbling around setting up the camera. The magic is in the moment, so have your camera ready and film the best you can with what you have...which leads to the next filming tip....

• **Put effort into your shots.** Even though they might just look like shoot-from-the hip style shots, most of the impactful Insta360 videos or photos you see took effort. Select a few mounts for each activity and keep it simple, but once you start filming, put your energy into getting "the shot", not just a bunch of average shots. This may mean you need to walk closer to your subject, or pause a moment longer, but the images you create will reflect your energy. There is a huge difference between "I could have taken that shot" and "I took that shot." With the latter, the proof is in the image.

MOUNTING ELEMENTS

Next, let's cover the basic mounting elements used for mounting the Insta360 X3 camera.

TRIPOD MOUNT (1/4"-20)

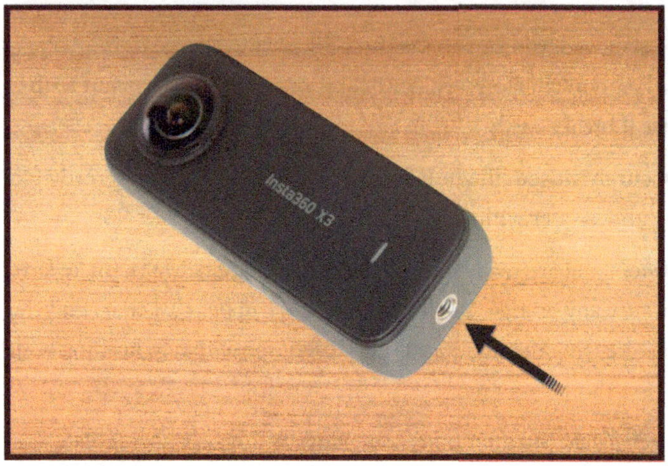

The Insta360 X3 comes with a 1/4" tripod mount on the bottom of the camera allowing you to **mount it directly to a variety of mounts with a tripod stud.** This is simplest, most direct mounting option since no additional pieces are required to mount the camera.

- When mounting your camera directly to a tripod stud, make sure the camera is **screwed on tightly** to prevent it from coming off when bumped.

- The **second option** for attaching your X3 to a mount is to use a **GoPro-style mount**. using a mount that adds mounting fingers the bottom of the camera. The mount to use for this is **a "tripod mount adapter"** with a 1/4" stud on top and mounting fingers on the bottom of the mount.

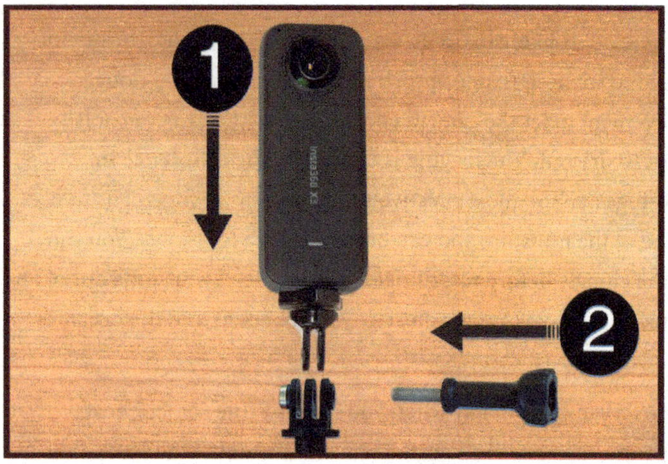

- The mounting fingers can be **attached directly to some GoPro-style mounts or used with GoPro mounting buckles.** A **long Thumbscrew is used to connect the mounting fingers to a mount.** Tighten the thumbscrew extra tight to prevent the camera from being wobbly at the mounting fingers.

HOW TO MOUNT THE X3 FOR THE INVISIBLE POLE EFFECT

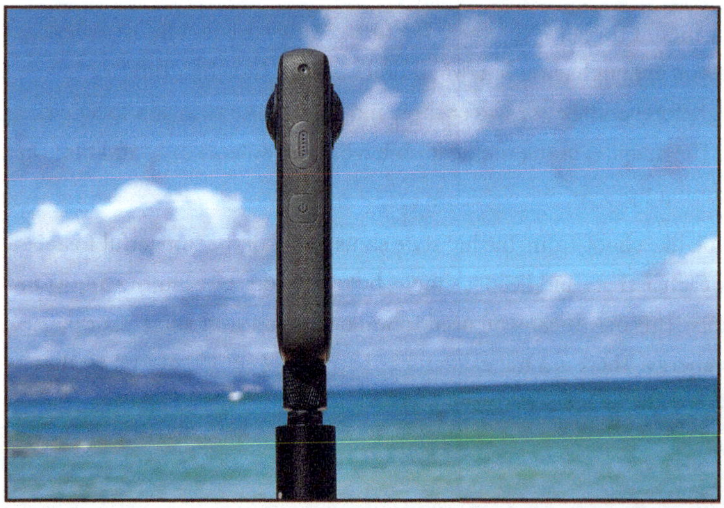

One of the best creative features of the Insta360 X3 is the ability to **make the pole disappear,** creating a floating camera affect. Whether you're a filming action, lifestyle/vlog, real estate, or scenery, the floating camera effect is definitely the greatest asset to using a 360 camera.

As long as your camera is mounted either **directly to a pole or to a "direct to camera" mount** and lined up with the pole, you can work the invisible pole for your 360 shots.

Notice how the camera is parallel to the pole, creating the invisible camera effect. The Insta360 poles are designed to disappear when the camera is mounted as shown.

• The way to do this is to mount the X3 on a straight pole like shown. For the best stitch lines, use a pole that places the camera at least 10 inches off a surface or away from your hand (if you are holding the pole). You want to use a pole that is about the same depth as the camera lenses or thinner (about 1-1/4" maximum pole diameter). Since each of the two lenses records more than 180 degrees, the overlapping areas in the stich line results in a disappearing pole.

• If you are using a GoPro-style mount, **a direct to camera mount**, like the one pictured to the right is the best option for mounting the X3. A pole that uses a "buckle" will not disappear completely from the shot and is not recommended. For 360 filming, use an extension or handheld pole and mount the camera straight in line with the pole.

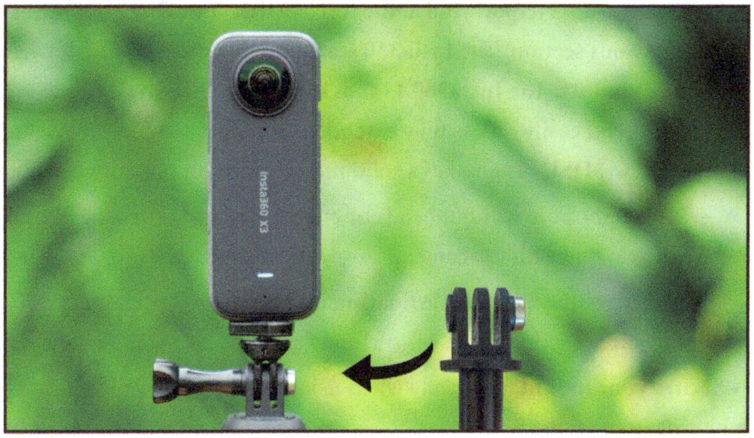

A direct to camera mount allows the camera to be mounted directly in line with the pole.

• For the best results when filming in 360, **the top mounting attachment needs to be centered** so the camera lines up directly with the pole.

• When recording Single Lens Mode videos, it's better to tilt the camera at the end of the pole to angle at yourself or your subject. The exception to this is when filming in Me Mode. In Me Mode, the camera has to be mounted parallel with the pole for the invisible pole effect.

BUCKLES

Some mounts, such as GoPro adhesive mounts, only work with buckles. If you decide to use a **"tripod to GoPro adapter" mount,** which is a great way to expand your mounting options, you can mount the X3 to a GoPro-style buckle. This buckle then locks into place into an adhesive mount for a secure way to mount the X3 to surfaces, such as a car, surfboard, boat, and more.

For shots recorded in Single Lens Mode, mounting the camera directly to a buckle is not a problem and can produce some amazing footage.

Unfortunately, when mounting the X3 directly to buckle mounts for 360 shots, the buckle does not disappear, resulting in an obvious stitch line and some strange distortion. Mounting the X3 directly to a buckle is not the ideal way to mount a 360 camera, but when used as the mounting base for an extension (see next session), this allows you to use the wide variety of mounts available to make the mount disappear. You will find that using **an extension pole on a buckle is one of the most creative ways to mount your 360 camera.**

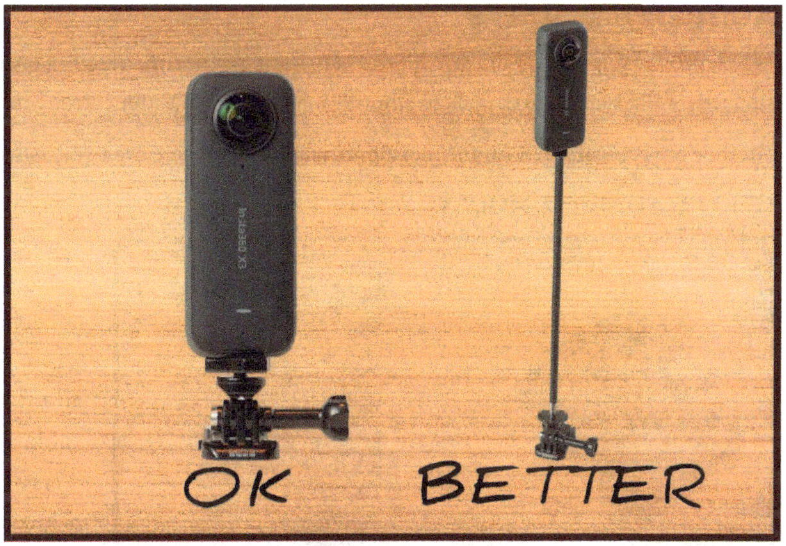

For 360 videos or photos, use an extension pole when mounting the Insta360 X3 to a buckle when possible.

• The **Insta360 Quick Release Mount** is also a great option which allows you to slide your camera out of the base quickly and easily. Also, there are lots of Quick Release options available from third party sellers, such as Ulanzi. Some of these are magnetic, while others work by rotating the mount in the base.

EXTENSION POLE

• You can **use an extension arm or section** with many of the mounts to capture unique perspectives and create the invisible mount effect. An extension arm is used to lift your camera out and away from a mount or your equipment. See the Unicorn Mount and the Mounted Extension Pole Section for examples.

• Use a pole with a **tripod stud** at the top or a **direct to GoPro mount** extension pole with a top connection like the one shown in the previous section. The **Insta360 Carbon Fiber Extension** (shown in the previous photo) is the **most sturdy extension for wearable mounts.** The **Insta360 114cm Invisible Selfie Stick** also works wonderfully as an extension since it is extendable, tough and won't bend easily unless fully extended. The Insta360 Invisible Selfie Stick can be connected to multiple segments using the tripod screws. **HSU** also makes a solid metal extension set that won't bend or flex.

• Use **a straight pole** without bends or curves. Bends or curves will cause the pole to appear in the image because the pole won't be directly in line with the camera.

• If you want to buy a **premade floating extension,** SP Gadgets makes all of the pieces to assemble your own floating extensions, including a 12" section and a 24" section.

• **Secure your mount,** especially when using an extension pole! Once you have figured out exactly how you want the camera pointing, tighten the screws well to secure your camera and the poles. The extra leverage of the pole makes it more susceptible to falling off.

• If you are using GoPro-style mounts, **use a Philips screwdriver to tighten any thumbscrews** that hold the camera in place. If you don't tighten the screws properly, the camera can accidentally rotate in the mount, and you won't know if you are still capturing the right angle. Even small bumps and shakes can move your camera around very easily, so use a screwdriver to tighten the thumbscrews as tight as you can.

• We will learn about some of the best ways to mount an extension pole later in this step.

BEST MOUNTS FOR 360

Now that you know about the individual mounting pieces and how to put them together, we will focus on the best mounting tools to use with the X3 in general and also the best mounts for capturing the "magic floating" camera shots in 360 photos and videos. **Let's take a look at the best mounts to use with the Insta360 X3!**

ADHESIVE MOUNTS

When properly applied, Adhesive Mounts are **the most secure base for mounting your camera or an extension pole to most surfaces.** Adhesive Mounts stick to smooth surfaces using a super strong adhesive tape. They are single use and can be removed, but they are **typically left on for more permanent use.** There are two basic adhesive mounts to choose from:

CURVED ADHESIVE MOUNT

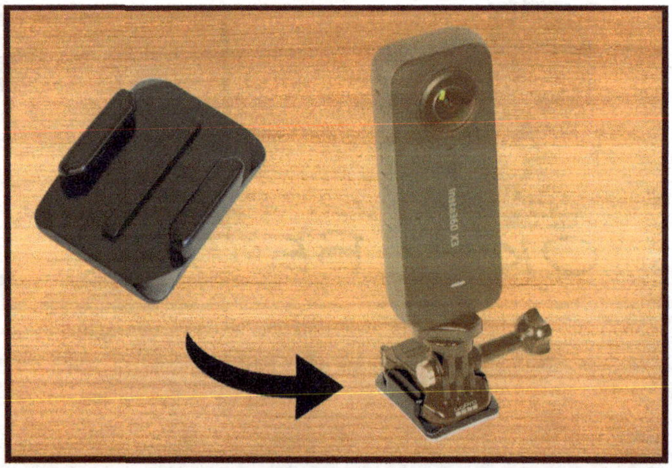

A Curved Adhesive Mount (aka Curved Base) has a slightly curved base **for mounting to rounded objects,** most commonly a helmet or a curved surface of a vehicle. Insta360 offers a curved base with some of their mounts, and GoPro makes a Curved Adhesive Mount that requires a buckle.

Mounting Tips:

- When using this mount on a helmet, find the best position and curve to match the base of the mount. Inspect all the edges of the mount to make sure you have a good fit. If the edges of the mount lift slightly because the curve of the mount doesn't match the curve of your helmet, you can secure it using a small amount of epoxy or sun-curing surfboard resin (like Solarez). Make sure to only apply the resin around the base of the mount, not on the part of the mount where the camera slides in.

- This is the best mounting base to use for mounting your X3 on a helmet. See the Action POV Mounts for more ideas on where to use this mount for Helmet Cam angles.

Mounting Example: THE "UNICORN" MOUNT ON A HELMET

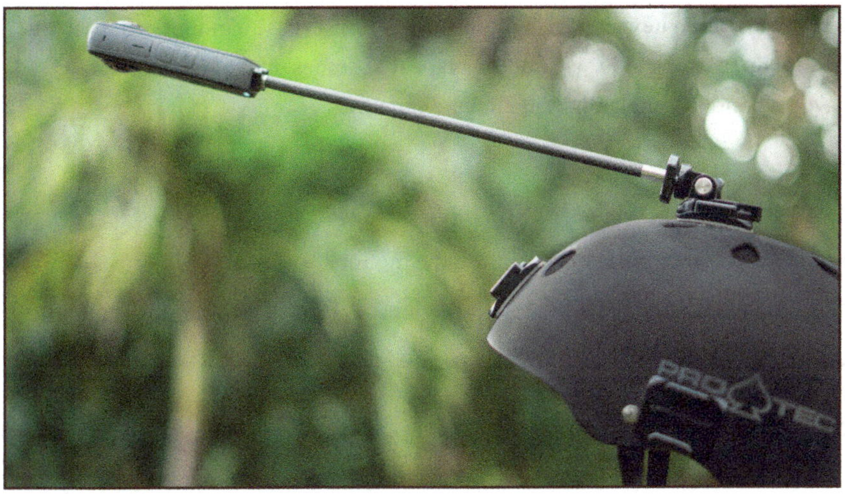

- RECOMMENDED SETTINGS: When using this mounting setup, record in 360 Video Mode at 5.7k-30 for high quality video with the invisible floating camera effect. If you are making lower resolution videos for mobile sharing, use 4k-60 since slow motion can help highlight the action with this type of shot.
- Also known as the Unicorn Helmet Mount, the Insta360 X3 is mounted on the end of the Insta360 Carbon Fiber Pole. The extension is mounted to Curved Adhesive Mount. All thumbscrews were tightened using a screwdriver to prevent them from moving.
- Although the weight of the mount is noticeable, the distance from the helmet creates an amazing floating camera effect in front of the subject.

FLAT ADHESIVE MOUNT

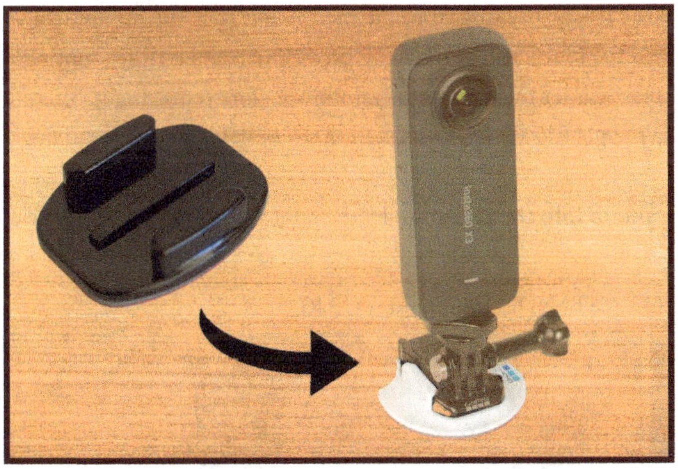

A Flat Adhesive Mount is a simple, low-profile mount **for flat, non-flexible surfaces.** The Flat Mount has a flat base and rounded corners. When mounted properly to a flat surface, this mount provides a secure, semi-permanent base to hold your camera. The best application for a Flat Adhesive Mount is to a location where you can use the mount repeatedly- e.g. your own car, plane, boat, or sporting equipment.

Mounting Tips:

- Follow the Adhesive Mount Mounting Instructions at the end of the Adhesive Mount Section to set up a secure mount. The mount is very unlikely to fall off if mounted as instructed on a non-flexible surface.

- **Drilled Mount.** For a secure mount in applications where a typical adhesive mount might come off, you can **modify the mount so that it can be drilled into a surface.** To do this: 1) Drill a hole in the middle of the mount, 2) Countersink a hole for the head of the screw (so it doesn't interfere with your camera sliding on) and 3) Screw or bolt your mount onto your board, surface, etc.

- The **GoPro Surfboard Mount** (shown above) is a flat adhesive mount with **more surface area** than the standard Flat Adhesive Mount. It comes with a separate tether mount that fits together snugly with the cutout on the Surfboard Mount. The extra surface area creates a stronger bond and can be used for other mounting setups besides just on a surfboard.

• Do not use an Adhesive Mount on a **SoftTop surfboard or bodyboard.** If you are riding a SoftTop surfboard or a bodyboard, **use the GoPro Bodyboard Mount** instead.

More Tips For Using Adhesive Mounts:

It is especially important to follow the mounting instructions if you are using the mount in cold weather. The following tips will help you use the Adhesive Mounts to their full potential.

• **Do not apply the mounts to a flexible surface,** like the nose of a snowboard. The mount can come off when the surface flexes.

• To **remove a mount** from a tough surface, you can usually pry it off with a butter knife. If you are worried about damaging the surface, use a hairdryer to soften the adhesive and slowly peel it off.

• When setting up a new angle, the best option is to **always use a new adhesive mount.** If you must reuse a mount, make sure to use a new piece of adhesive tape. **3M VHB 4991** is the super strong adhesive bonding tape that comes on GoPro® mounts. You can pick up a roll online and it's always good to have on hand if you plan to reuse any mounts.

• When possible, use a camera tether for backup.

• Adhesive mounts can withstand temperatures up to 250ºF (121ºC).

• When mounting your camera, make sure the buckle "clicks" into place to lock it into the mount.

• **ADHESIVE MOUNT MOUNTING INSTRUCTIONS:** Follow these steps to securely mount your camera using an adhesive mount:

1. First check to **make sure you are using the right adhesive mount** for the surface you are mounting to. If you are mounting to a flat surface, use a Flat Adhesive Mount (Base) or Surfboard Mount. When mounting to a curved surface, use a Curved Adhesive Mount (Base). Before removing the backing from the adhesive, test the mounting position to make sure the edges of the mount sit flush on the surface.
2. Use isopropyl alcohol to **clean the surface** where you are going to place the mount. Make sure there is no wax or sand on the surface. Let the alcohol dry or wipe clean before moving to the next step.
3. If you are using a GoPro base, decide which direction you want the camera to face when mounted and **make sure the groove for the camera to slide into faces that direction**. You can insert the camera forwards or backwards, but you can't easily rotate the camera once it is mounted.
4. The mount works best if **applied at room temperature.** Peel the plastic backing from the mount and stick the mount to the surface. NOTE: The adhesive does not feel extremely sticky to the touch and needs to be pressed hard onto the mounting surface. For tricky surfaces, use a hair dryer to heat up the red plastic liner that covers the adhesive before removing it. This will make the adhesive tackier and improve the adhesion. Some people use a lighter to soften the adhesive, but be careful not to burn it.
5. For the strongest bond, **wait 24 hours before placing your camera into the mount** to let the adhesive form a strong bond. After 72 hours, the mount will be fully set to the surface.

Mounting Example: ADHESIVE MOUNT ON THE NOSE OF A STANDUP PADDLE BOARD

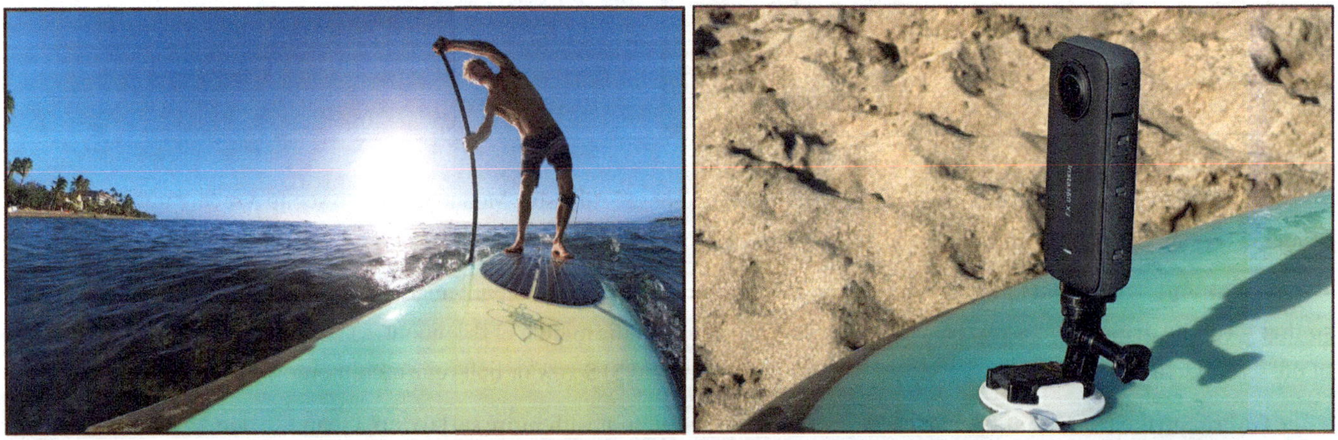

• SETTINGS USED: Video recorded in 360 Video Mode at 5.7k-30 for reframing to multiple perspectives from one clip. In general, when using this mounting setup, record in 360 Video Mode at 5.7k-30 for high quality video. If you are making lower

resolution videos for mobile sharing, use 4k-60 since slow motion can help highlight the action with this type of shot. If the surface will be moving a lot, you can alternatively use Single Lens Video Mode at 2.7k-60 with Horizon Lock Turned on (using FOV+), but the view is more limited.

• The X3 was mounted directly to the GoPro Surfboard Flat Adhesive Mount.

HANDHELD POLE

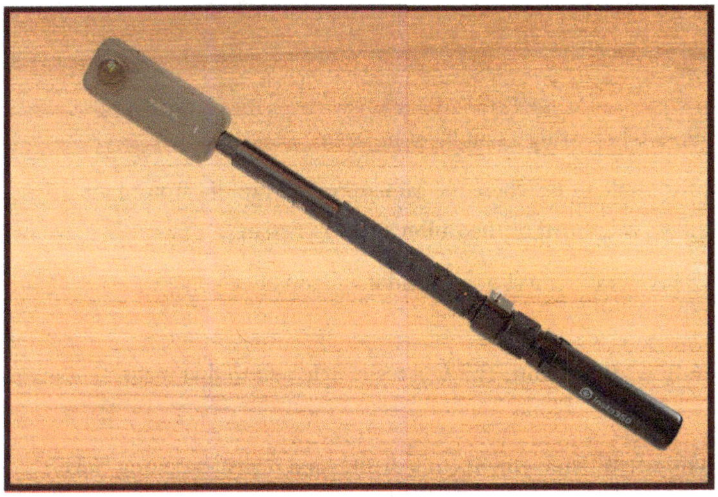

When filming 360 photos and video, hand-held shots give you **the most flexibility for unique angles and creative camera movements.** And since the pole is not visible, all of these movements happen magically, as if someone else is filming.

A handle (typically about 6"-8") or a pole (18" +) provides an easy way to hold your camera and **capture multiple angles with one mount.** A handle or pole (also known as a selfie stick) enables you to capture some of the best angles possible for any activity where you have a free hand. The pole creates distance between you and your camera for ultra-wide-angle shots with lots of scenery or for close-up reframed videos. You can **film yourself, follow behind your friends, or hold the pole vertically** to get more height on your shots.

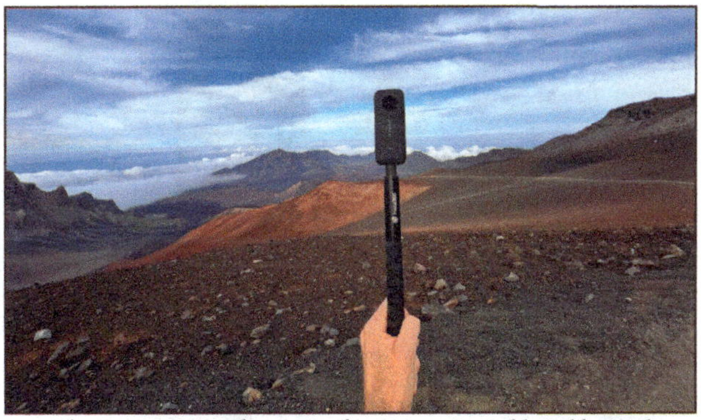

The Insta360 X3 shown on the 114cm Invisible Selfie Stick.

If you want to hold your camera on a pole (but look like you are not), the **Insta360 Invisible Selfie Sticks** are the most obvious choices. These extendable poles have a tripod stud attachment and are designed to disappear when used with the X3 (due to the width of the poles), giving you lots of flexibility with a relatively compact size. There are a few sizes available, but the 114cm pole is small, versatile, and still provides a lot of extension.

An additional addon to the selfie stick are tripod feet so you can easily set your camera down on the go, extend the pole and start recording. The **Insta360 Bullet Time Handle** is one of the best options since it also has a tripod stud on the side for creative Bullet Time shots. The Bullet Time Handle screws directly into the base of the Invisible Selfie Sticks of the Extension Pole. Insta360 also makes the Selfie Stick with Tripod 2-in-1 pole, which is another great option.

When using any extendable selfie stick for selfie shots, **extend the pole out at least two feet from yourself for the best stitch lines and perspective** on your spherical imagery.

For a longer handheld pole, **270Pro** makes a high quality, extendable pole which extends up to 270cm (106 inches). With the wide-angle lens on the X3, the super long pole really makes it look like the camera is far from your subject or even flying through the air for **fake drone shots**. Made out of carbon fiber, this pole is light and won't break on you, unlike some of the cheaper plastic options. Because of the loose sections, this pole can also be used for the trendy **"Javelin" shots,** where you extend the pole out quickly into a tiny planet view.

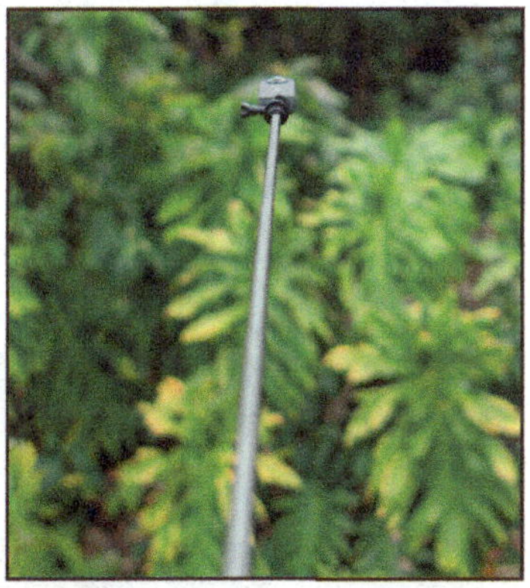

The super long 270Pro Extension pole can be used to create an aerial view.

The **270Pro Backpack Pole** offers five different extension heights, which is much more convenient than the regular pole, which only offers two. Insta360 makes a similar high quality pole, called the **Insta360 Extended Edition Selfie Stick,** which is also made out of carbon fiber and aluminum and extends to 9.8ft (117cm). If you are into extreme sports, adventure, or just like to hang your camera off of high cliffs for unique shots, you want to have a pole you can trust.

For a pocket-sized handle, the **GoPro Shorty** has a direct to GoPro mount and provides just enough extension (8.9") to move the camera away from your hand. Plus, it has foldable tripod legs for quick and easy tripod shots on the go.

If you are going in the water with your camera, it's best to find **a pole that floats.** SP Gadgets makes a 12" floating extension that mounts on the end of a floating handle. The pieces connect using a 1/4"-20 screw, so there are no mounting pieces to get in the way of your invisible selfie stick shots.

More Tips for Using a Handheld Pole:

• See the examples for the **two primary ways to hold your camera** when using a handle or polecam.

• As you learned in Step 2, Bullet Time Shots are creative handheld pole shots where the camera rotates rapidly around the subject. You can best create these using either the **Insta360 Bullet Time Cord** or the **Bullet Time Accessory.**

• **Hold the pole in front of you or behind you** while you ride a bike, walk, travel or surf and it will look like you have a personal film crew following close by.

• When filming with a handheld pole, **move the camera so it follows the action.** This will really help add to that magic floating camera effect.

• If you are using a super long pole for fake drone shots, try to **mimic the movements of a drone** for a realistic looking "fake" drone shot.

Mounting Examples

VERTICAL FILMING

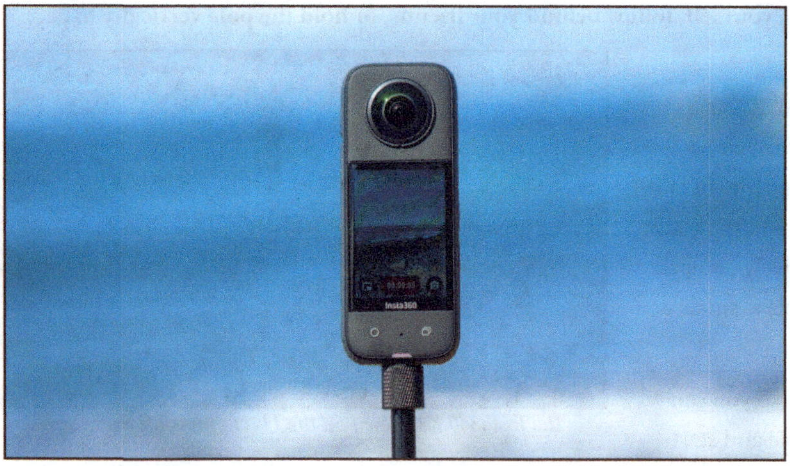

• Hold your camera in a vertical position **when you want to add more height to your shots.** When using the X3 with a long pole, a little bit of height goes a long way. The ultra wide-angle lenses make the camera appear much higher than it actually is.

• A vertical position also works well **for vlog style shots to center yourself in the frame.** When you go to reframe your shots, having yourself or your subject closer to the center of the frame will result in less distortion.

HORIZONTAL

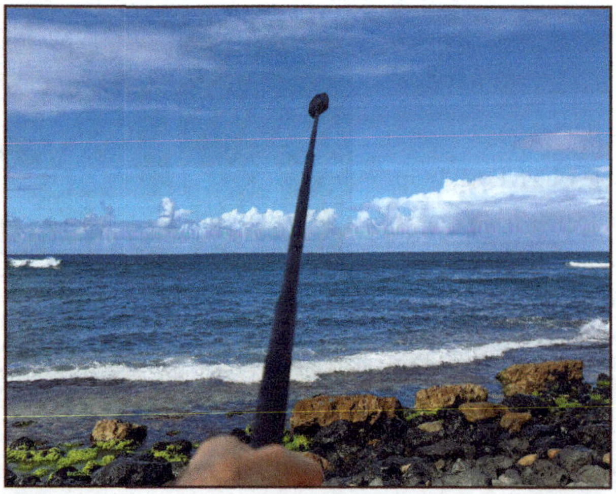

• Holding your camera horizontally on a pole is better when you are trying **to create a sense of moving along with yourself or your subject.** This filming position places the camera more at hip or shoulder level for an immersive ride along.

• For anyone with a traditional photography background, it feels wrong to not point the camera lens towards yourself or your subject. However, because you are filming a full spherical image, this is **one of the best techniques for the invisible camera effect.**

• If you are filming **in Single Lens Video Mode, angle the camera at the end of the pole** so one of the lenses (the one you choose to film with of course) is facing your subject.

MONOPOD

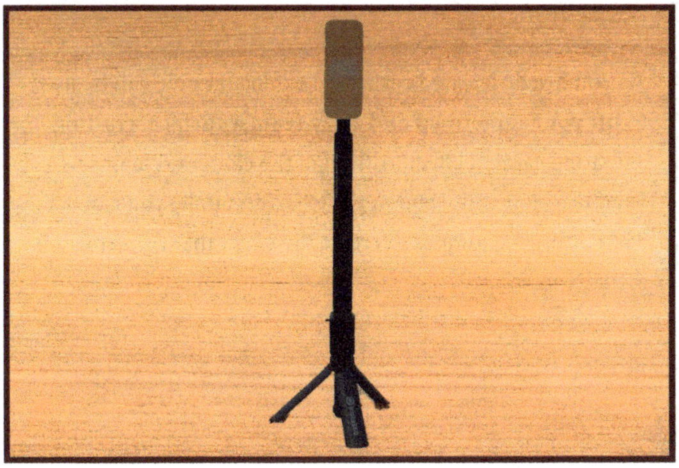

Using a monopod (essentially a selfie stick with tripod feet) is the **best option when you want to set the X3 down for a stationary shot**. The long pole leading down to the monopod feet disappears easily out of 360 shots, leaving a minimal amount of touch up work to edit out the tripod feet. A tripod on the other hand will result in a lot more postproduction editing.

A monopod provides the perfect mounting base to set the X3 down, press the Shutter Button or use the Self-Timer and disappear from the scene (go hide behind a bush, vehicle or sign). For the most versatility, **look for a monopod that raises up to at least eye level.** If you do end up venturing into real estate 360 photography or virtual tours, that eye-level height will produce the most relatable imagery.

When choosing a monopod, also make sure that the adjustments (knobs, etc.) and pole diameter are **less than the width of the Insta360 X3 (about 1-1/4")** so the pole disappears in 360 images. If you have a pole with a ¼"-20 attachment at the base, you can just add a monopod base for an instant monopod. Zeadio makes a simple metal tripod base that attaches with the ¼"-20 screw attachment.

More Tips for Using a Monopod

- Be careful of filming on a monopod in **windy conditions.** A monopod is not as stable as a tripod and can easily get blown over in the wind. Since your lenses are not protected, this could be the end of your camera.

- If you are **filming 360 real estate photography,** a monopod is essential for filming all of your panoramic photos to create virtual tours. We will talk more about using the X3 for real estate in Step 4.

- Without the appearance of a pole in the shot, a 360 camera is an invaluable tool for someone filming travel, lifestyle, or vlog content. **Using a handheld pole with a combination of a monopod is the perfect combination** for recording a variety of shots without looking like you're filming yourself. You can pull multiple angles out of a single moment in time, allowing you to create fast, high-quality content on the go. When filming in 360, you won't forget to film a thing. In Step 4, you will learn more tips for filming vlog and lifestyle shots using the X3.

- A smaller monopod, at least 12" off of the ground, can be useful for mounting the X3 on a table or desk **for filming at home studio shots and online content.**

- If you are filming in sand, dirt, or grass, **a spike base** works even better so you don't have to edit out the monopod feet later. It's easy to make one of your own for this purpose by gluing together a PVC pipe, a GoPro Mount for the top connection and a plastic solar light spike for the base.

Mounting Example: MOUNTED ON THE INSTA360 114CM INVISIBLE SELFIE STICK

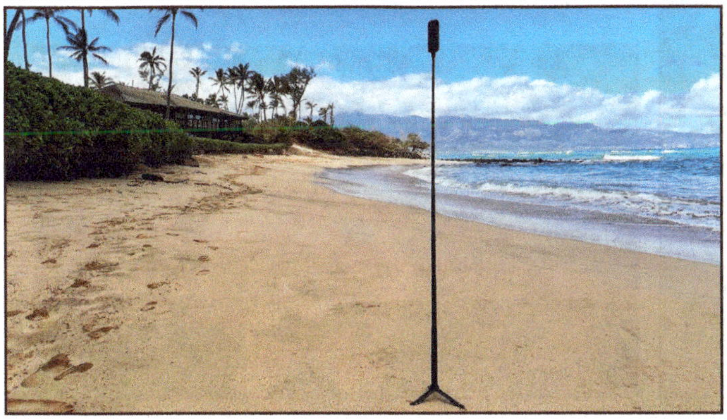

- The Insta360 X3 is shown here mounted on the 114cm Invisible Selfie Stick fully extended with the Zeadio Tripod Base. Although the pole will disappear in a 360 shot, the monopod feet will need to be edited out for spherical viewing. You will learn how to edit out the tripod feet in Step 5- Editing.

MOUNTED EXTENSION POLE

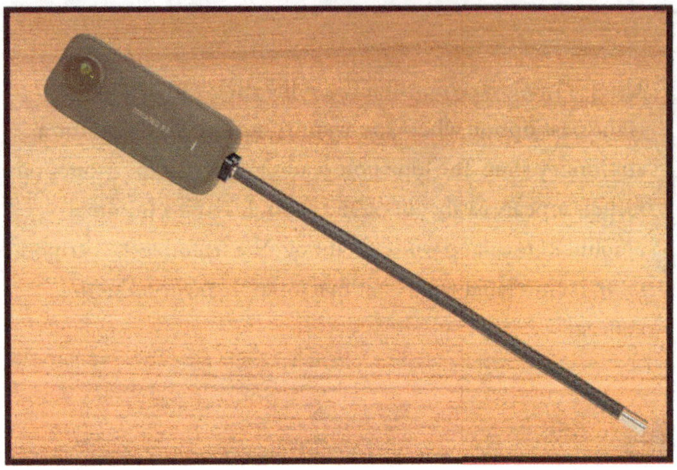

As you briefly learned earlier, an extension pole can be used to **lift your camera up and away from a surface,** creating a floating camera effect. Although the camera remains in a fixed position at the end of the extension pole, you can use 360 editing techniques to create the effect that the camera is actually moving.

This simple setup, a mounting base, an extension pole, and the Insta360 X3, opens up the potential to easily create shots that would have been extremely difficult previously. For example, imagine a shiny new car driving along a scenic cliff. With the Insta360 X3, mounted off the front of the car, you could easily make a cinematic car commercial with a relatively cheap and easy setup. The possibilities are in your hands!

There are several options available that make the best bases for mounting an extension pole:

- Use an **adhesive mount**, like we talked about earlier, for a more permanent mounting position on your equipment. Having an adhesive mount ready to go makes it easy to attach an extension when you're ready to start filming. The **Insta360 Carbon Fiber Mounts** are tough and ready to handle the X3.
- Attach a **Curved Adhesive Mount to the top of a helmet,** add your extension pole so it's facing out in front of you, and capture an amazing angle. This is called a **"Unicorn" or "Gnarwhal" Mount** (like we showed you before) because it looks like a narwhal's horn. Just be sure to tighten the base of the extension mount well so it doesn't rotate as you ride.
- For a more temporary mount, use Insta360's **Suction Cup Mount.** The Suction Cup Mount can easily be attached and detached from a variety of surfaces, such as a car or window. The Suction Cup is a versatile mount to use as you travel for quick, creative shots. The Suction Cup can be used on smooth, nonflexible, non-porous surfaces. Always test the Suction Cup's suction before you attach your camera. Add the extension pole to the Suction Cup Mount, and now your camera can appear to be magically floating off of any surface.
- **Insta360's Third Person Bike Mount or GoPro's Handlebar Seatpost Mount** are two useful options for attaching your camera to a pole. These mounts attach securely around a pole, which serves as a solid mounting base that won't come off.
- For a super strong magnetic mount, the **Bushman Popeye Magnetic Mount** holds up to 99 pounds of pull and is ideal for attaching an extension pole to any metal surface, such as the exterior of a vehicle.

Mounting Example:

EXTENSION MOUNTED TO A GUITAR USING A CLAMP MOUNT

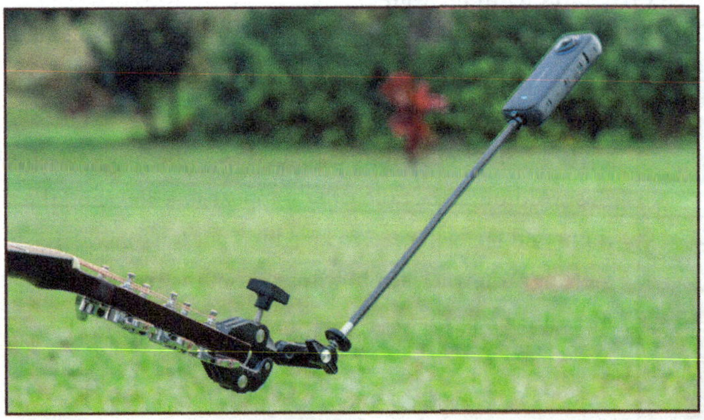

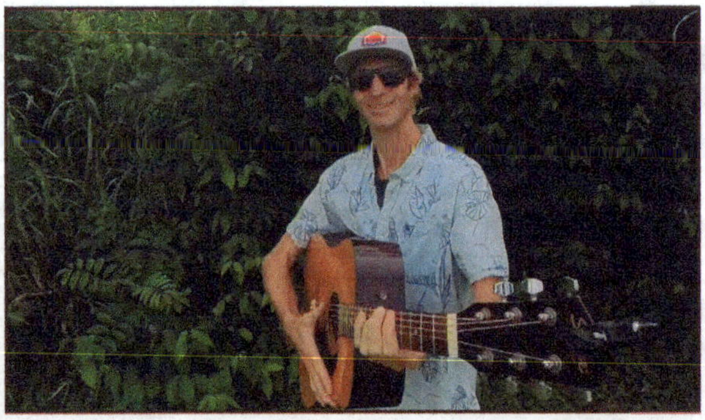

• RECOMMENDED SETTINGS: When using this mounting setup, record in 360 Video Mode at 5.7k-30 for high quality video with the invisible floating camera effect. By zooming out when you edit, you can create the effect that the camera is mounted much further from the instrument than it actually is.

• The X3 was mounted to the Insta360 Carbon Rod to the Ulanzi Clamp Mount.

ACTION POV MOUNTS

Point of View (POV) shots **show the action from a first person perspective**, as if the viewer is doing the activity. This style of mounting is filmed by mounting your camera to your body or your equipment. Although many of these mounts don't allow the invisible mount technique (at least as dramatically), the super wide angle possibilities of the X3 make the first person view even more dramatic than even the wide angle lens on a traditional action camera.

Here are a few of the best first person mounts for filming 360 spherical shots (and of course Single Lens Videos too):

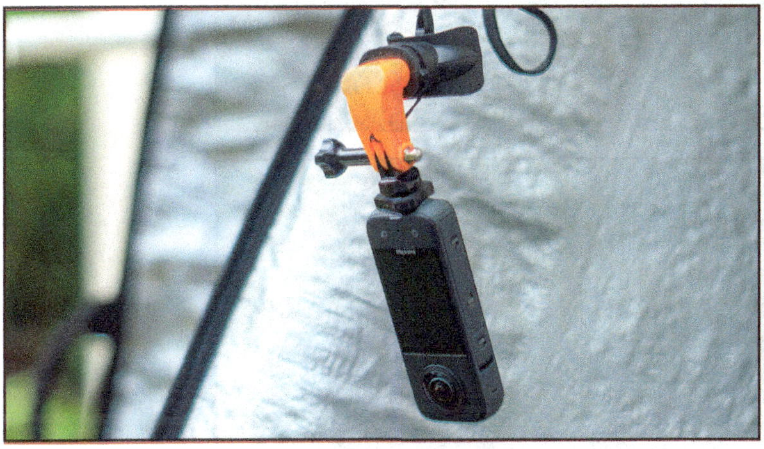

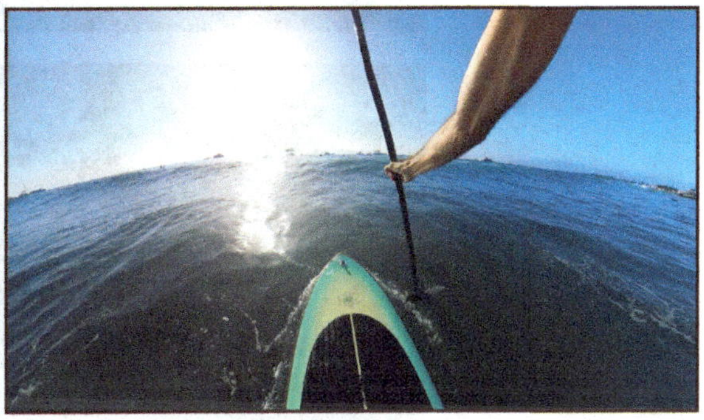

• **Bite Mount.** Popularized by surfers who had few other options for a good POV mounting position, people of all disciplines have found out how useful this mount can be. When using a bite mount, you actually hold the camera in your mouth using a mouthpiece. Some bite mounts, such as the MyGo Mount, have a similar mouthpiece to a snorkel, while others, like GoPro's Bite Mount, is more of a flat plate. A bite mount seems strange at first, but it really creates the most similar perspective to an eye level view. When filming in 360, the forward facing view can be zoomed out for a super wide angle, especially if you are reframing to 16:9 (because of the wide side-to-side perspective. The result is a super dramatic shot because you can see more of the surrounding scene. If you zoom out too far, the view through the lens facing the subject is horrible because it's so close, but this can be carefully edited to only include the forward-facing view.

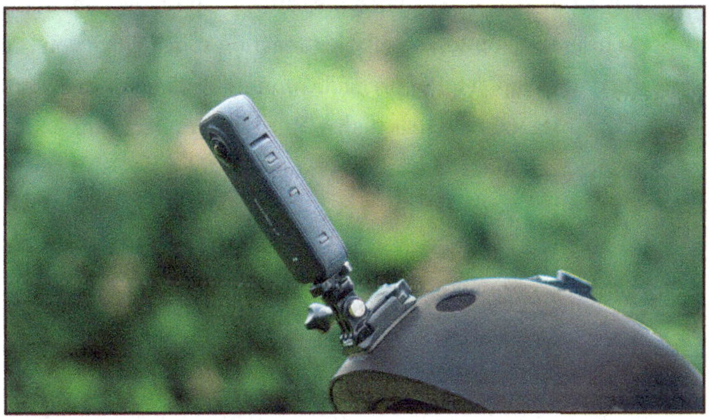

• **Helmet Mount.** Mount the X3 directly to a helmet for sports where you wear one. The X3 can be mounted **facing forward on the front** (just above the rim) for the best forward facing view. Mounting the X3 directly **on top of a helmet** allows you to see all the way around-forwards, backwards, and on the sides. If you add an extension pole to either mounting position, you can also capture that invisible mount perspective with a POV view, which is extra intriguing.

• **Chest Mount.** A chest mount can be a great place to mount your X3 for a first person point of view. With the camera mounted pointing straight out from your chest, reframe the shot to an ultra wide angle perspective until your arms come into view, allowing your viewers to see your movements.

• **On your equipment.** As we mentioned in the section on extension poles, you can mount the X3 directly to your equipment using an extension and an adhesive mount. This perspective can create some amazing follow-along footage and is great for action shots. You can also use some of the other mounts we talked about, such as Adhesive Mounts, the Suction Cup, the Third Person Bike Handlebar Mount, or the Handlebar Seatpost Mount, to mount your camera to your gear, depending on what surface you are mounting to.

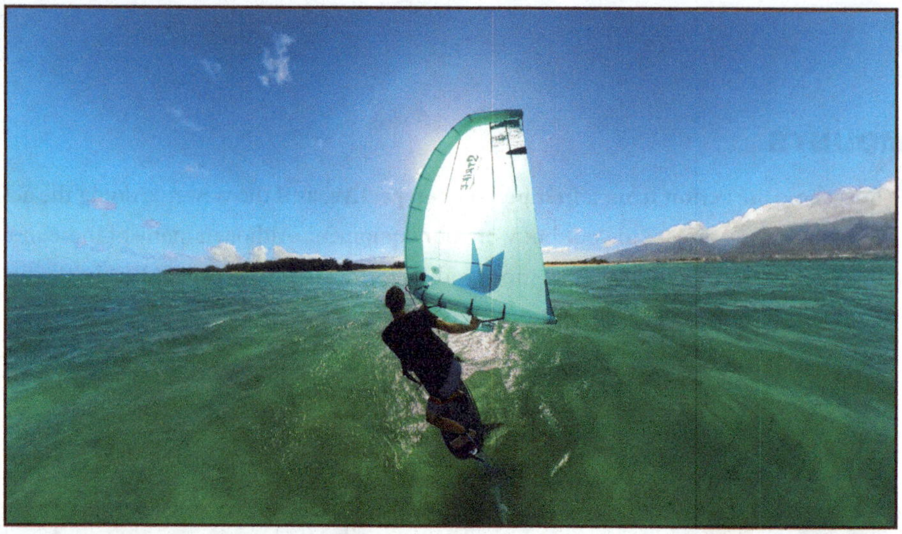

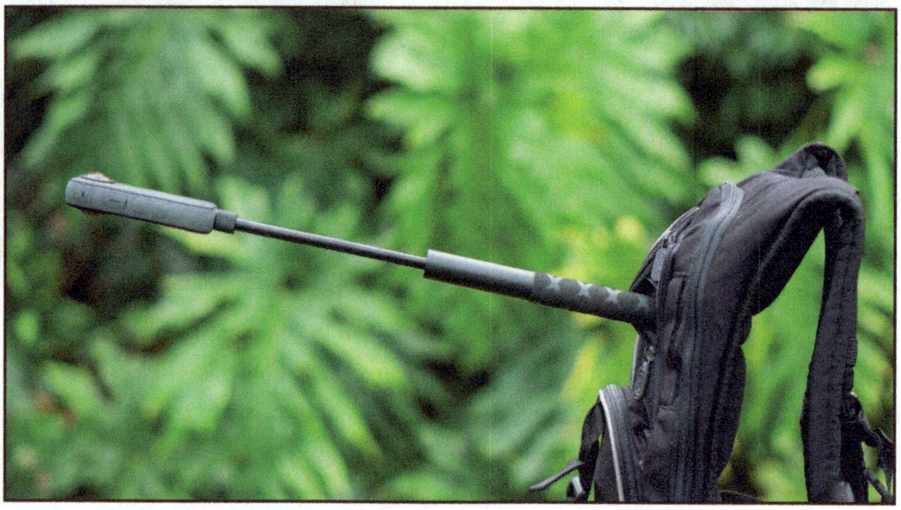

The Insta360 114cm Invisible Selfie Stick Mounted to the Muvi Backpack.

• **POV Harness or Pole Backpack.** A POV Harness or a backpack with a pole mount allows you to mount the Insta360 X3 out and away from your body so that the camera can follow your action while the mounting set up remains a mystery. This is one of the best mounting options available for action sports since you can really follow the action without seeing the pole. For a long time, there were not a lot of premade mounts available with a waist harness or backpack and a straight pole like you need for 360, but luckily Insta360 now sells both a Third Person Backpack Mount and the Back Bar (a waist mount). Muvi also makes a third person backpack mount like the one shown.

More Tips for Using a POV Mount

• In Single Lens Mode, the Horizon Lock feature is a huge bonus when filming POV shots. As you move and rotate, your camera can keep a steady view of the horizon, resulting in super smooth action shots. This creates amazingly steady videos when mounted to a bike, snowboard, child or pet.

Now that you've got your camera mounted, you are ready to move onto Step 4 to learn some vital photography and cinematography knowledge!

STEP FOUR

CAPTURE YOUR ACTION

Learn These "Not-So-Secret" 360 Video and Photo Secrets & Get Results

Now that you understand the Insta360 X3's settings and have discovered the best mounts to maximize the magic of this camera, it's time to start capturing life's moments. If the setup was the lead in, this is the climax of the story. With the filming tips and tricks in this step, creating a photo or video that holds as much energy as life itself will become natural.

We will break this step down into three sections- filming in 360, working with the X3 camera, and specific X3 filming tips for vlogging, real estate, action sports, and travel. These techniques will take your filming from standard to extraordinary, so let's begin!

360 FILMING TIPS

Filming or photographing in 360 has its own unique set of skills. Some elements of what's important in traditional photography remain and some are thrown out the window. So, what do you need to think about when composing your shots in 360? **These tips will help you with the most important elements:**

STORYTELLING

When planning shots for your video, remember that you want to tell a story. If you want to make a video that grabs your viewers' attention, **start thinking about your video before you are out there in the action.** The best action videos inspire us and motivate us to go adventure by telling a story, showing the action, and leaving on a high note.

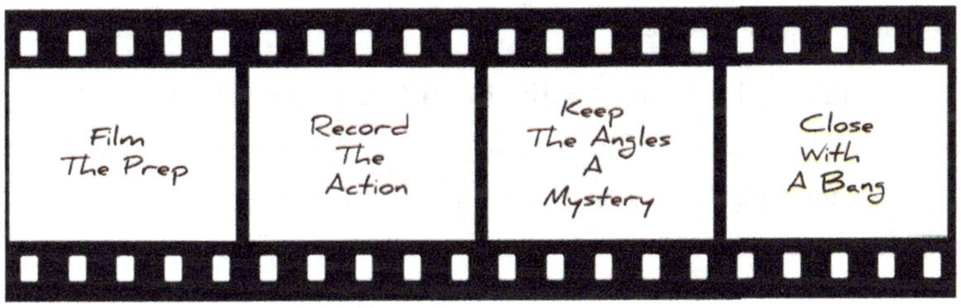

FILM THE PREP

Instead of just recording the main activity, film the preparation. Place your camera close when putting on your gear or prepping your important equipment. On your way out, film a pulled back shot (either on a monopod or a follow shot) leaving for your destination. If filming in 360, as you film, imagine how you will edit these shots later. In Single Lens Mode, change settings as you film. Use the Linear lens for a more traditional look.

RECORD THE ACTION

This is where you can use your mounting knowledge to record the action. This is where you **show off your skills and most memorable moments with high action shots or beautiful scenery.** Record in hi-res 360 for the most editing freedom when you edit. If you are planning to share to social media and want slow motion, use 4k-60 in 360 video mode. In Single Lens Video mode, film at 3.7k-60 in ActionView 2.7k FOV+ using the Max Lens for immersive action shots.

MAKE THE ANGLES A MYSTERY

When possible, try to **film multiple angles of the same action without seeing any other cameras** in the scene. This keeps the viewer involved in the action, not in the filming techniques. This is where you can apply your newly acquired mounting knowledge. When filming in 360, you can also pull multiple angles from the same shot, facing forward and at you, for example.

CLOSE WITH A BANG

Close your video with some humor or a dramatic point of view shot that leaves your viewer wanting more, or better yet, wanting to go out there and do what you are doing- having fun!

LIGHTING

As with any photography or cinematography, filming in **quality lighting is always going to give you a better starting point** for your imagery.

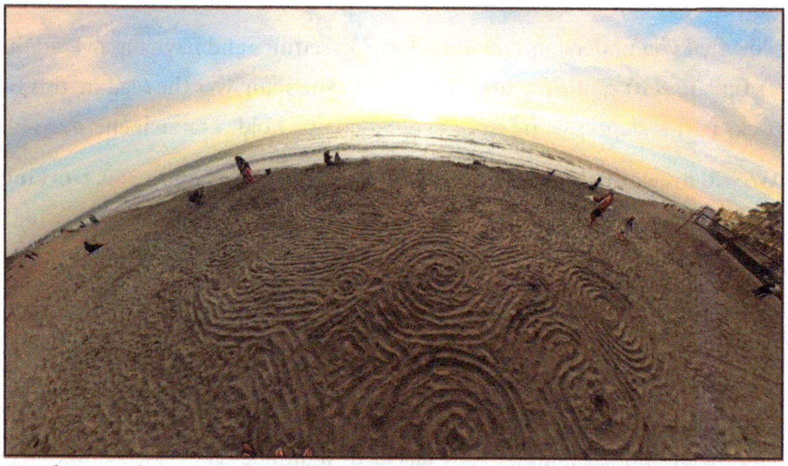

For the most dramatic lighting, film around sunset or sunrise, known as the Magic Hour in cinematography. Photo taken from a video recorded in 360 Video Mode at 5.7k-30 with Max ISO set to 400.

Of course, you can film anytime of the day, but morning and afternoon lighting produces richer colors. The hours around sunrise and sunset, known as **the Magic Hour or Golden Hour,** are definitely the most dramatic. The long shadows and rich colors during these hours stand out, especially when viewed side-by-side with a mid-day clip. Whenever you can capture photos and videos at these picturesque times, go for it. The resulting footage will make you look like an instant pro.

That being said, a lot of life happens between sunrise and sunset. **Filming in shade,** especially solid shade, also works well for avoiding harsh midday shadows.

Sunny, blue skies or scattered clouds create much more appealing, color rich shots than overcast days. Overcast conditions typically create monochromatic, dull colors and washed out scenes. If you have to film in overcast conditions, try to minimize the amount of sky in your shots by choosing covered locations. Also, editing your footage into black and white can play off of these monochromatic tones and add drama to dull footage.

Another common technique for portraits and fashion is to **film backlit shots** of your subjects. With the sun at your subject's back, typically only soft shadows remain on his or her face. The trick with these shots is that they need to be exposed for the subject, not the overall scene. You can also use Exposure Value Compensation in the settings set at +2. This will make the camera overexpose the shots slightly. You will probably need to fine tune the final exposure when you edit.

TIP: **Split Lighting.** When recording in 360 with the X3, pay attention to how the lighting is different on each side of the camera. Since both lenses are recording, try to avoid bright light on one lens and dim light on the other. With the X3, the best way to film in split lighting conditions, such as around sunset, is to **rotate the camera so both lenses share the lighting equally.** Alternatively, when you know you will primarily be using the footage from one of the lenses (say your subject is in front of the sun for example), use Isolated Exposure in the settings to set the exposure for each lens independently. That way you can avoid placing your subject on the side of the frame and possibly near the stitch lines.

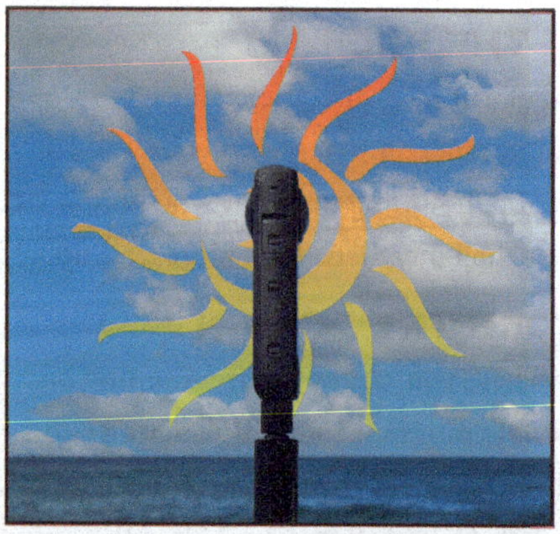

TIP: Low Light Situations. Filming in low light situations (early morning, evening, cloudy or shady), will often create blurry photos or videos.

When filming handheld shots in low light, around sunrise or sunset, **raise the ISO to 800** or even to 1600 depending how dim the lighting is.

For the best results when you are recording video **in low light situations, use a video frame rate of 24 or 30 frames per second.** A lower frame rate allows your camera to take in more light for each frame, resulting in better quality footage. A higher frame rate (60FPS) is better suited for normal daylight. You can also **raise the ISO anywhere from 1600-3200,** but remember that a higher ISO results in video with more noise.

TIP: Speed Blur/Motion Blur. You can **use the low light to your advantage** to create a cool, speed blur background effect. A speed blur usually has an object of focus, surrounded by scenery that appears blurred. In bright daylight, the shutter time will naturally be quick: this creates a shot where everything is in focus and appears frozen. That's perfect for shots when you want that look. This look can, however, take away from the action of the moment and make the scene look stagnant. A slower shutter doesn't freeze the action as much, and as long as you have an object of focus (usually your bike, car, or whatever your camera is mounted to), the speed blur reflects the real movement of the scene.

The easiest way to create motion blur is to **mount your camera to a moving object in low light.** The object will stay in focus because it's moving with the camera while everything passing by will appear to be a bit blurred. If you want to create more blur, lower the ISO to 100. For less motion blur, raise the ISO.

For example, mount your camera to the front of a bike facing back at you so you and your camera move in unison. In low light situations, the camera will not have enough light to keep the passing objects in focus, giving your photos or videos the blurred background effect. This technique works specifically well and looks really eye-catching for Interval photos but is also noticeable as motion blur in regular speed video shots.

You can also **add motion blur in Insta360 Studio** when you speed up your videos into TimeShift videos.

CINEMATIC 360 TIPS

One of the biggest questions new Insta360 users want to know is, "How can I make my videos look like a motion picture?" Filming in full spherical 360 is definitely a unique experience, but that doesn't mean that the results of your edited video need to be any less cinematic. With all of the creative editing possibilities, you will find that it's actually easier to create cinematic videos with a 360 camera.

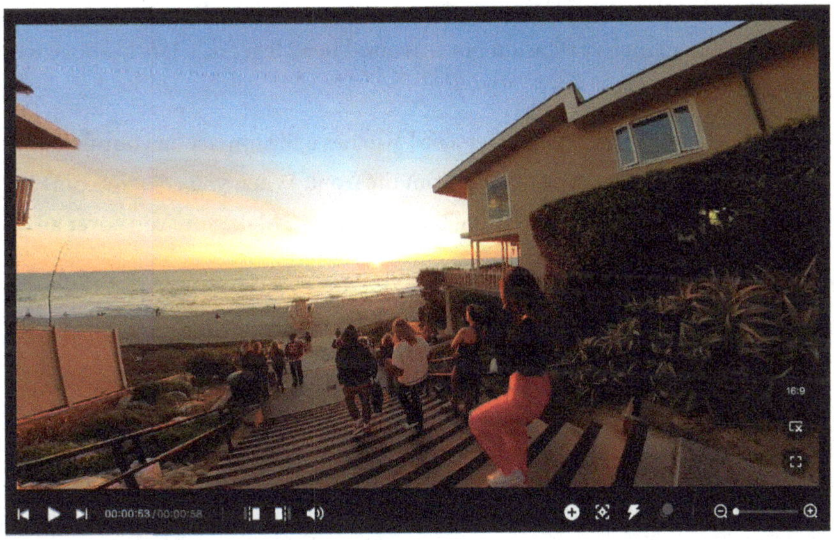

With the right settings and techniques, your Insta360 footage can easily mimic cinema quality films.

Follow these tips to give your video footage a more cinematic head start. The final cinematic results will happen in the editing room:

- Take advantage of the **highest quality settings** on your camera. Film in **5.7k at 24 or 30 FPS** and let this be your go-to. If you reframe to 1080p, this gives you lots of extra resolution for creative editing. Also use a low ISO (100 preferably, 400 max) for bright light shots to get the best quality noise-free video and photos.
- Utilize **slow motion along with movement.** The filming techniques shown in this step, such as panning and tilting, look great with slow motion. Filming at 60 FPS opens up the possibilities for slow motion. In 360, you can also use the ultra-slow motion technique you will learn to create super slow motion from your 5.7k-30 spherical videos. FlowState Stabilization gives you the creative freedom to film these shots even without a gimbal for smooth footage. If it's bumpy, it won't look cinematic.
- **Put energy into your filming.** Take your filming techniques serious and film as if you are using a bigger cinematic camera. Use a tripod if you need to and treat your camera like it's a cinema camera.
- **Film when the light is right.** Lighting makes the mood and can't be faked even with great editing. Film during the **golden hour** for mood and dramatic shadows. Use **light leaks** to let light filter in and out of your scene. Playing with natural light will make your shots look magical.
- **Film transitions.** As you film, try to remember to film a few of the creative transitions you will learn in this step.
- Film in Vivid Color for the richest color without much color correction required in editing. When editing, add in **contrast** if needed to deepen the blacks. **Adjust colors** to get your desired look. LOG color will need color correction to get that cinematic look. Consider exporting your video at 24 frames per second for the most cinematic feel.
- Use **cinematic fonts** to overlay titles in Premiere Pro, DaVinci Resolve or another editing app. Or add a logo to your intro for a professional look. There are tons of fonts available, but a few to get started that can be downloaded for free are Montserrat, Eurostile, Helvetica Neue, Bank Gothic, Couture, Nexa and Bebas Neue.

PERSPECTIVE AND ANGLE

- **Use perspective.** Adjusting the height of your camera in relation to your subject makes a big difference in the perspective of your shots. Some shots look great when your camera is closer to the ground. Shoot from **hip-level** for a more childlike perspective. Use a pole to place your X3 **really high** for an aerial view. Place the camera **close to the ground** for a mouse's view. The variety will help to add a sense of wonder and uniqueness to your shots. The height from which you record creates a perspective, and even though you can reframe your shots when you edit, you can't adjust that perspective.
- Especially because of the super wide angle, a 360 shot filmed with your camera set down on the ground is going to look a lot different than a shot recorded high up on an extension pole. **Around head level or slightly above will give you the best perspective for people or vlog shots.**
- **The angle** (on the side, behind, in front) from which you record also has a big impact on your final shot. Following behind your subject is going to produce a far different image than filming from the front. When you edit, you can't magically rotate the camera around your subject unless you actually filmed the shot using that movement. Make sure you cover all of the angles when you are out there filming.

CAMERA MOVEMENT

Filmmakers use a lot of different movement techniques to make their videos look more cinematic. Some of these movements pertain to filming 360 videos, while others are done when you edit. You will see how easy it is to create some of these movements when you learn to edit in Step 5- Creation.

It's not really necessary for a recreational filmmaker to know the names of the various filming techniques, but an awareness of the various techniques will inspire you to add more flair to your videos. When you watch videos that impress you, pay attention to techniques the filmmaker used to make the video stand out.

In Step 5, you will also learn how to add similar movements to your photos in a snap.

Since most recreational 360 camera users don't have expensive filming equipment, it's best to **make do with what you already have** to mimic these techniques in the grassroots do-it-yourself action cam style everyone loves. The amazing FlowState Stabilization on the Insta360 X3 makes all of these shots really smooth, eliminating the need for a tripod or gimbal in most situations.

TILT SHOT

This technique is like **looking up and then down** or vice versa. A tilt shot is filmed by pointing the camera up and then down by rotating it on its horizontal axis. This shot is typically filmed using a tripod.

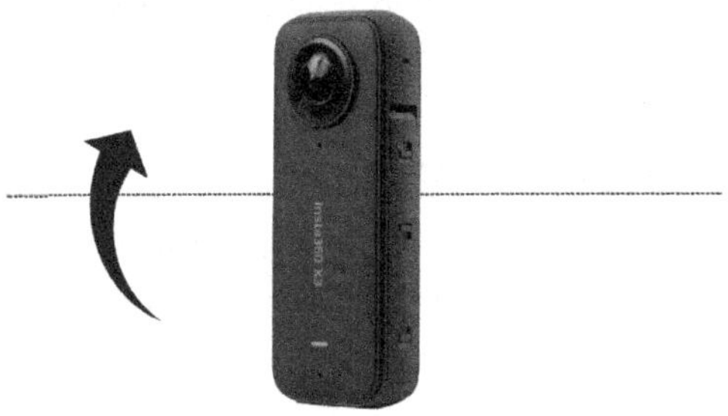

But with the X3, since you have a full sphere to work with when editing a spherical video, you can **mimic a tilt shot** when you edit even if your camera was stationary, and in reality, it is best to leave your camera still when possible.

PAN SHOT

A pan shot is similar to a tilt shot, except, instead of moving the camera up and down, a pan shot is **filmed in a side-to-side motion**, rotating the camera on its vertical axis from left to right or vice versa. This technique is like looking left and then right.

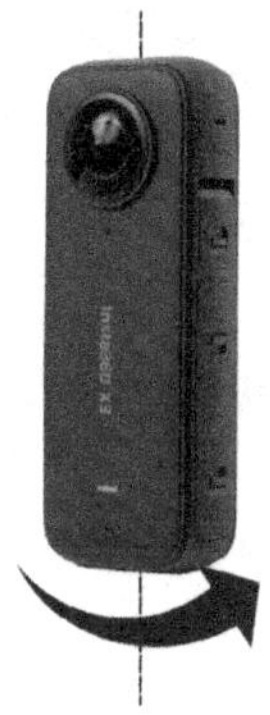

Once again, this is one of the cinematic filming techniques that can be replicated when you edit your 360 videos. For eye-catching shots, rotate your camera from an empty scenic shot towards your subject, bringing your subject into the frame. For best results, don't edit the movement in your shots to pan too quickly- too quick of movement can cause jittery effects.

DOLLY

A dolly shot is typically filmed with the camera mounted on a camera dolly, moving the **camera towards or away from the subject.**

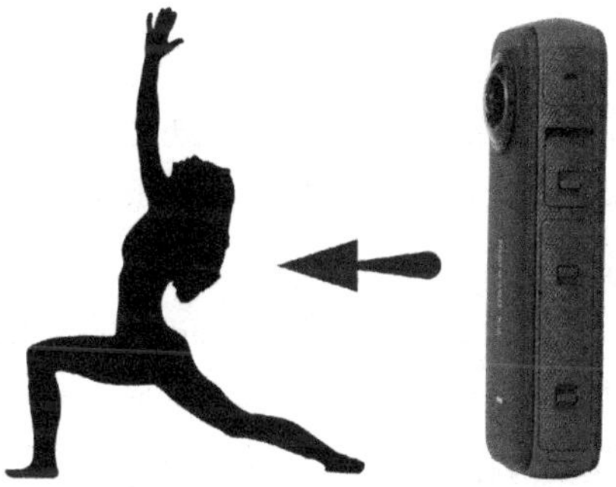

In-camera stabilization makes this shot really easy without the use of a dolly. Film your dolly shots by holding the X3 steady while moving your camera slowly towards or away from your subject. This filming technique looks great with stationary subjects for intros to a video.

To some extent, you can also edit your reframed 360 videos to mimic this filming movement, however, you will be limited to how far you can move in on your subject while maintaining high quality video.

PEDESTAL

A pedestal shot involves **physically moving your camera up or down** in relation to your subject. Since this actually changes the perspective of your subject, this one is better filmed on the spot. Trying to recreate this when you edit looks more like a pan shot than a pedestal.

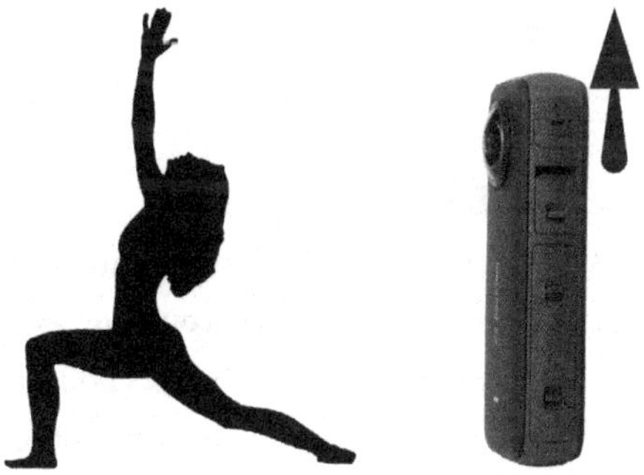

TRACKING SHOT

In a tracking shot, the camera moves from left to right or vice versa, keeping the camera on the same axis to **move parallel with the subject.**

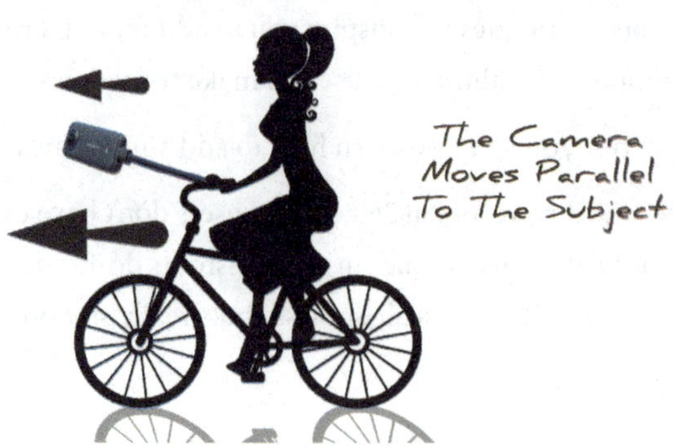

This is one of the cinematic movements you will really want to capture as you film. A great way to mimic this technique with the X3 is to hold the camera on a pole while riding a bike or skateboarding. The goal is to create smooth movement that moves with yourself or a subject and this happens naturally when holing the X3 on a pole. When possible, include passing foreground objects between you and the subject for a real sense of movement.

FOLLOW SHOT

In a follow shot, the camera **physically follows the subject at a (more or less) constant distance so you will want to actually film this movement, rather than edit it in later.**

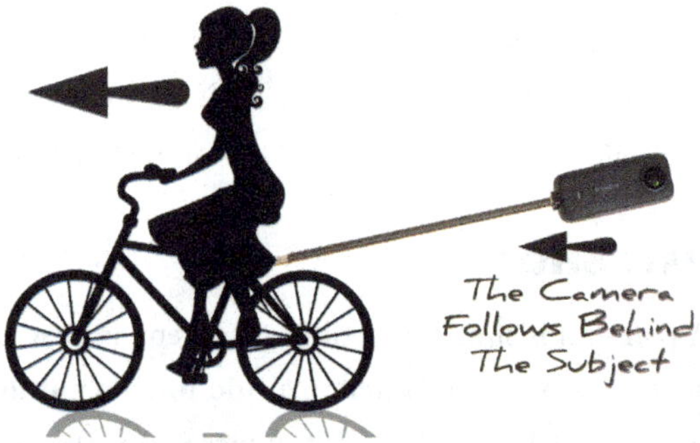

You can capture this angle of yourself by mounting the X3 on an extension to the back of your equipment, vehicle, etc. With FlowState Stabilization, you can also ride behind someone on a skateboard, bike, snowboard or surfboard to follow a friend.

PASS THROUGH

Pass through shots are really fun to add into your clips. To film a pass through shot, utilize the invisible camera effect to physically **move your camera through a space** (such as a tunnel or a tree stump) or between objects (such as stair railings, pier pylons or a window). When you edit the 360 video, you will have lots of fun reframing the journey. When filming with the app, you can use **Fly Through Mode** by pressing the paper airplane icon on the right side of the screen in 360 Video Mode. This will give you prompts to create pass through shots that will be auto-edited by the app, which makes it a lot easier for you!

6 IN-CAMERA VIDEO TRANSITIONS

When it comes to editing, most editing apps offer a variety of preset transitions to create move smoothly from one video clip to the next, but you can really set your footage apart by using the X3's stabilization to **film in-camera transitions out in the field.**

Typically, in-camera transitions are filmed using a gimbal to create smooth movements, however by utilizing the in-camera stabilization in the X3, you can really master transitions without the need for a gimbal. If you film with these transitions in mind, when you edit your clips, the hard part is already done. It just comes down to editing two clips together to make the transition happen.

> **TIP:** When editing, use the Cross Dissolve/Crossfade transition in an advanced editing app to transition from one scene to the next (like we will show you in Step 5.)
>
> In-camera transitions are used add more flair to your quality video- kind of like a topping on your ice cream. Pick your favorites and mix them in at the beginning and end of a few of your key shots.

STRAFE BLOCK

This is probably the easiest transition to film and actually remember on the scene. For a strafe block transition, **move your camera sideways using a foreground object close to the camera to blur** the end of one scene and transition to another. By filling your frame with an object (someone's body, a vehicle, a board, etc.), you can then use the same object to transition into the next clip. The key is having the same object at the next scene.

With the ultra-wide angle lens on the X3, you will need to get as close to the object you are filming as possible to completely cover the frame. When you edit, you will have to remember the transition and reframe the shot to create the transition. Alternatively, you can use Linear or a zoomed in field of view in Single Lens Mode to reduce the wide angle perspective.

PUSH IN/ PULL OUT

Move in on a subject to **completely cover the frame,** and then pull out to reveal a new subject. For example, you could walk towards your subject, filling the frame with your subject's back. Then move locations and start your shot close to your subject. Pull away to reveal a new scene.

The push/pull transition is similar to the strafe block, except that it defines the camera's movement as moving in and out on a subject. Once again, you won't be able to completely fill a 360 sphere, but you can compose this shot in the Touch Screen and edit it to fill the frame when you reframe the video.

THROUGH THE GLASS

Creating the effect that you are **moving from outside of a window through to the inside** is easy using a 360 camera. Just use the Suction Cup to film a view outside of a window or door in 360. Then, film the inside view from the opposite side of the door. When you edit, it's easy to rotate the shot to make it appear that the camera moved right through the window or door.

SKYFALL

Start on a subject. At the end of your shot, **tilt up to the sky.** Start your next shot aimed up at the sky. When you tilt back down, bring the viewers into another clip or scene. This transition is easy to create with the X3 because you always have sky in your 360 shots.

WHIP-PAN

When filming in low light, **utilize a motion blur** to end one scene and begin another. Create a fast side-to-side motion to end a scene. In the next scene, continue that motion so you can fade in to begin the next clip.

THE HAND PUSH

As you film, **add in a few hand gestures** as if you are pushing your camera to the side or pointing in a certain direction. When you go to edit, these hand gestures can be used to quickly reframe your shot to the new perspective. For example, if you are walking into a museum filming yourself, point to the museum, then reframe your shot to show your viewers what the museum looks like.

SETTING UP YOUR 360 SHOTS

A lot of the shots you will be filming with the X3 will probably be used for reframing into a traditional format video or photo, which means you will zoom in or out, add keyframes and flatten the view when you export.

However, if you want to record full spherical 360 **with the end goal of sharing a 360 video or photo,** there are some **special filming tips** you should follow for the most mesmerizing 360 videos and photos possible:

Scope The Scene- Take the extra time to look around to find the best vantage point for filming a full spherical view. By centrally placing your camera in a scene, your audience will have more opportunities to look around and explore. Also, take into account the audio because spatial 360 audio adds in another element of wonder. If you get the audio right, it will add a lot to your 360 videos.

Disappear From The Shot- If you are filming a 360 scene, it's best to start recording and then hide out of view. For photos, you can use the Self-Timer to hide from view or the Insta360 App/ Remote to press the Shutter remotely. A 360 video or photo is much more mysterious without the filmmaker standing by. Since you won't be by the camera, make sure the camera is in a stable position that won't be blown over by a strong gust of wind and that it's also in a safe spot where no one will take it.

Watch Using the Insta360 App- If you are close enough, you can preview the scene, start recording and watch a live view using the Insta360 App. You can also take 360 photos remotely using the App.

Give It Time- Depending on your goal for the video, you will want to video at least a few minutes. This will give you enough time to reframe the video for different view, or for a viewer to look around the scene.

UNIQUE 360 SHOTS

There are some shots that are totally unique to 360 spherical photography and cinematography that are good to know about when you are out filming. Let's explore how to capture these unique 360-specific shots. You will learn how to edit these in reframing tips in Step 5.

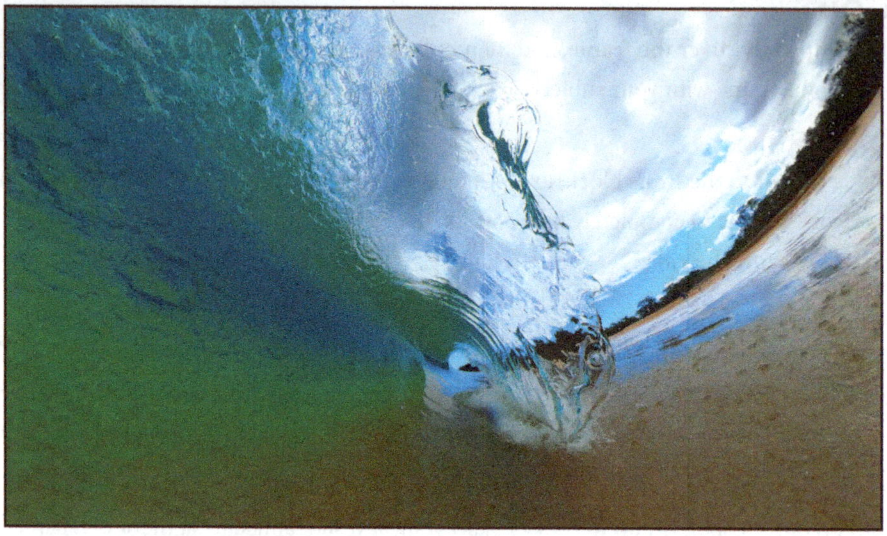

A Super Wide 360 Photo captures a wider perspective than is possible with any traditional camera. Edited in Insta360 Studio.

Tiny Planet or Rabbit Hole. A tiny planet shot is where the photo or video is zoomed all the way out, making it appear that the subject is on a tiny planet floating in the sky. The opposite of a Tiny Planet is called a Rabbit Hole shot, where the sky is in the center of the shot and the landscape swirls around it. Both of these shots can basically be reframed from any video or photo filmed in a 360 mode. You can create tiny planet videos, photos, or timelapses, or simply use a tiny planet perspective as one of the views in a reframed video. To film a tiny planet shot, mount your camera on the end of a handheld pole, monopod, or extension pole so there is plenty of distance between your camera and the surface it's sitting on (or from the subject). High cloud formations and sunset colors will help add more drama to these shots.

Super Wide. A 360 camera like the X3 is capable of producing super wide angle shots beyond what a single lens can capture. By pulling out on the shot, you can create unique perspective that were previously impossible. Super Wide shots look amazing as scenic panoramas and action POV shots.

Running The Planet. (Roll Planet) A Roll Planet shot is a Tiny Planet or Rabbit Hole shot with a subject running around the sphere in the middle. To appear that you are running around the tiny planet, set the camera on the ground and move in a circular motion around the camera.

Jump Planet. A jump planet is also a type of tiny planet shot, but his one is created by mixing together different scenes to make it appear like you are jumping planets. To film a jump planet shot, jump up when holding the selfie pole in variety of scenes.

Fake Drone. Fake drone shots are easy to film with a long extension pole. Raise your camera into the air and move your camera around like you are flying a drone. When you reframe the shots, it's easy to use a pulled back perspective to make the camera appear to be even higher than it was.

Javelin Shot. Mount your camera on the 270 Pro Pole with the pole compressed. Move forward and fling the pole up, extending all of the extensions. When you edit, move to a tiny planet at the end of the shot for a great effect.

TRICK SHOTS (SHOT LABS)

The mobile Insta360 App offers automatic editing for a wide range of trick shots, some of which would be very difficult to edit manually. These trick shots are located **in the Shot Labs section** under Stories in the mobile app. To give your own footage the special editing effects in these trick shots, all you need to do is open up a Shot Lab and follow the instructions in the app. The hardest part about using Shot Labs is remembering to record specifically for one of the shots.

Take a look through the Shot Labs in the app and **see which ones you like.** Then, when you go out filming, take this book with you when you go out to shoot as a **reminder of which Shot Lab shot you want** to make your own.

SHOT LABS

Unless noted otherwise, **use 360 Video Mode at 5.7k-30** for the following Shot Labs:

<u>With the Pole Held Vertical</u>

CineLapse- Record multiple clips of different scenes.

Fly Lapse- Recording while walking for 3 minutes.

Jump Planet- Hold pole vertical. Jump, wait 2 seconds and jump again in multiple scenes.

<u>Hold the Selfie Stick Naturally</u>

Sky Swap- Record a scene with open sky and walk.

Auto Time Shift- Record for 3+ minutes as you walk/run.

<u>Hold Selfie Stick To the Side at Waist Level</u>

Stop Motion Mix- Walk for 40 seconds. Record at least 3 scenes.

Stop Motion- Record 2 minutes of video.

Time Flip- With your back against a wall, move the camera in a semi-circle from one wall to the other side. Record multiple scenes, moving the camera the same direction.

Split Jump- (Use 3k-100 Video) Rotate pole backwards as you jump in multiple scenes.

<u>Hold Selfie Stick Behind You Over Your Shoulder</u>

Flash Dash and **Horizon Flip**- Walk, run, or ride for at least 1 minute while recording video.

<u>2 Person Shots</u>

CenterStage- Film your subject moving the camera position up/down/side to side. Hold for 1 second at each spot.

Flip My Day- Rotate Selfie Stick from parallel to the floor to vertical filming a subject. Repeat for multiple scenes.

Dolly Zoom- Record a 6-8 second clip walking towards or away from a centered subject.

Face Off- Stand face-to-face 1.5m apart with tripod between you, and position the lens facing each person.

<u>Tripod Shots (Mount your camera on a tripod/monopod)</u>

Shadow Clone- Perform different actions in a circle around the camera, making sure not to overlap.

Clone Trail and **Pixelize**- Move around the camera at 6ft. (2m)

Clone Loop- Move around the camera at about 4ft. (1.5m)

Street Lapse- Record 2 clips of one scene at Night- one with lots of cars, one with the subject and few cars. Wear light colored clothing.

Roll Planet- Walk around the camera once.

Parallel Planet- Run or walk 2 laps around the camera 6ft. (2m) away

Giant Jump- Stand 2ft. (.8m) from the camera with the camera low to the ground. Jump, wait 2-3 sec., jump again. Repeat in 6 scenes.

<u>One of a Kind</u>

Nose Mode- Hold your camera in your mouth between teeth. Wave your arms and run.

Freeze Throw- Throw the camera up and catch it. Try to prevent the camera from rotating.

Overtaker- Use GPS with the App on your phone. Mount your camera and drive or ride.

<u>Other Modes (Use the Mode Specified)</u>

Star Lapse- Use Star Lapse Mode. Record 6+ Hours with the camera stationary.

Bullet Time Mix- Use Bullet Time Mode. Record at multiple locations with the camera above your head.

Spin View- Use 360 Photo Mode. Take a selfie shot with the camera out at a 45 degree angle.

Ghost Town- Use Time Lapse Mode. Record for more than 1 minute.

ACTION

Once you are ready to start recording some of life's moments, how you film will have a lot of impact on the overall quality of your videos and photos.

Use these tips when filming the action with your X3 to give your videos and photos an experienced look:

• When filming, remember to **keep your camera steady and let the action create the movement.** With video shots, a steady camera helps create footage that is easy on the eyes- too much erratic movement and it becomes hard to watch, even with the best stabilization.

• Recording in 360 can be confusing to compose when you think about it too much. To start out, you can **film as if you are filming through one lens**, looking at the Touch Screen to compose your shots as you film. You can always reframe that same perspective from the 360 clip, but once you begin to edit, you will see the potential of having the spherical image to work with.

• When taking photos, **a steady camera** helps ensure that your photos will come out "sharp" and in-focus.

• For video shots, it's best to either film **using one of the cinematic filming techniques** shown in this step **or use no movement at all** (keep the camera still). This will give your videos a more polished, professional look and be much more enticing to watch.

• **Short clips vs long.** For editing purposes, it's easier to **stop recording in between "action" moments.** If it's too hard to reach your camera or your activity doesn't allow it, you can film continuously. It just means you will be dealing with larger files and searching through longer clips of footage later to find those "WOW" moments.

WORKING WITH THE 360 X3 CAMERA

The tips in this section will help you work with the unique qualities of the Insta360 X3 camera.

THE WIDE ANGLE LENS(ES)

The Insta360 X3 uses two built-in Fisheye Wide Angle Lenses to record the full sphere required for 360 shots. What does this mean? A fisheye lens is **an ultra-wide-angle lens that allows more of the scene to be included in the frame.** Especially when fully zoomed out, fisheye lenses cause curvature around the edges of the frame.

The Fisheye Wide Angle Lenses are perfect for the Insta360 X3 camera for several reasons:

1. The camera **can be mounted extremely close to you** and still capture you in the image.

2. The lens has a short depth of field, which means **everything from about 3-4 inches away from the camera and beyond will be in focus.** This is great because you don't ever have to focus the camera.

It is important to understand the best techniques for working with these fisheye lenses because wide angle photography is an art in itself. **Here are some vital tips:**

1. **GET CLOSE!** The fisheye lenses used on the Insta360 X3 make objects appear much further than they really are. If you aren't close to whatever you are filming, your subject will be very small in the frame. You know how some rear-view mirrors on cars say, "Objects May Be Closer Than They Appear". The same is true with fisheye lenses because they distort the perspective to make things look smaller than they look with the naked eye. Remember this and **get closer to your subject than you think you would need to.** By getting close to your subjects, your capture more engaging photos and videos.

2. When you edit, **placing the subject or horizon in the middle of the frame results in the least amount of distortion.** As you will see when you edit, as you tilt up and down, the horizon will become more curved and distorted. Because of this, when possible, try to place yourself or your subject closer to the center of the frame. It just makes reframing easier.

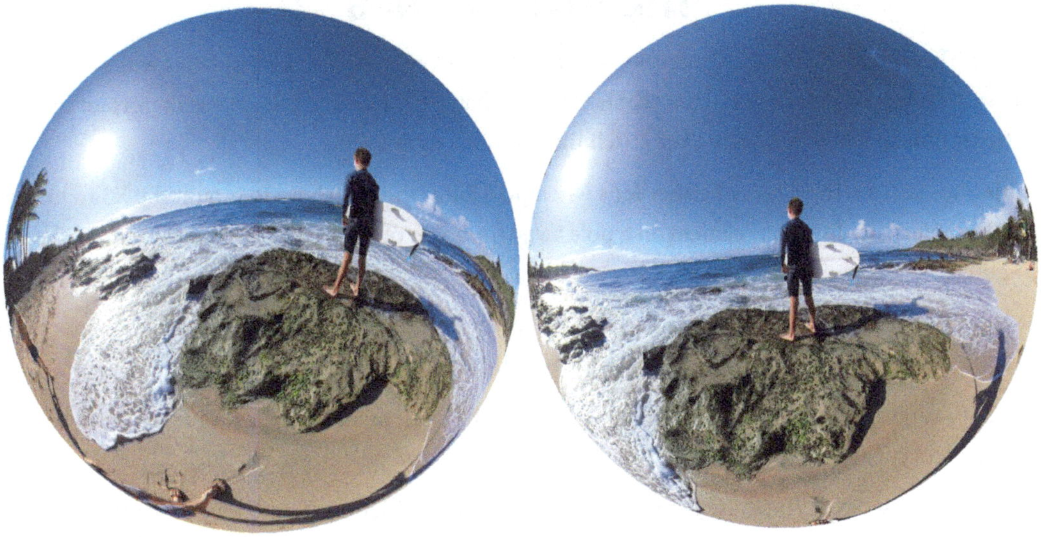

You can see the difference reframing can make in the appearance of your shots. Both frame grabs are from the same moment, but in the shot on the right, the centered horizon reduces distortion dramatically.

If the horizon is towards the top or bottom of the frame, it will have a curved appearance. When you learn to reframe your videos in Step 5, you will see the benefits of straightening the horizon.

Also, if you are holding the camera on a pole, more distance between the camera and yourself (or your subject) results in less of a fisheye bubble. This makes it easier to reframe the photos or videos without distortion.

In Single Lens Video Mode, you can use the Linear lens to reduce the fisheye effect.

3. **Use the foreground to your advantage.**

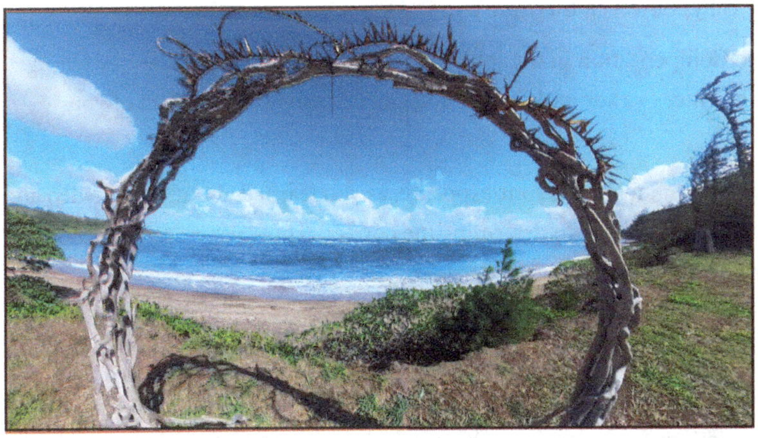

With the close focal range of the Insta360 X3's wide angle lenses, **give your image more depth** by framing an object near the foreground. Try placing the camera near some flowers, trees or your gear to create a sense of perspective. You can even create a cool look by "framing" your image with a foreground object. These foreground objects can add depth to both 360 and traditional photos and videos.

The natural art on the shoreline provided a great frame for the photo.

4. **Work with the sky.**

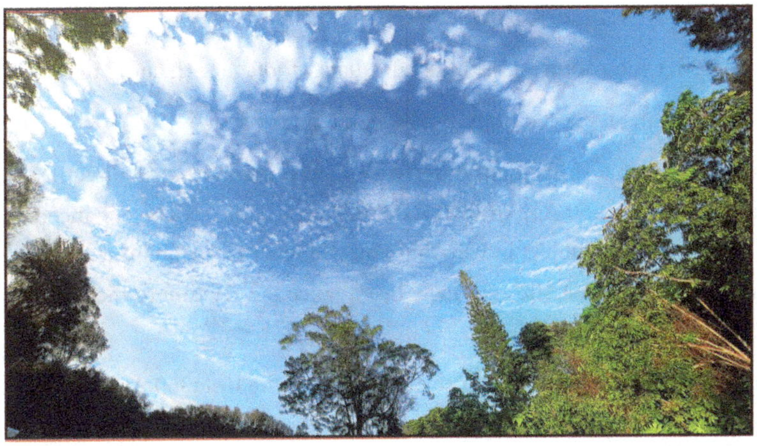

Because wide angle lenses capture so much in the frame, high clouds and sunset colors in the sky add a lot of drama to a scene.

The high clouds add to the composition, rather than detracting from it.

THE STITCH LINES

When filming in 360, The Insta360 X3 is pretty accurate at seamlessly merging the two hemispheres without a noticeable stitch line. However, the stitch line is sometimes partially visible. Making that stitch line unnoticeable is important to the quality of your 360 shots. **These tips will help you avoid seeing the stitch line as much as possible:**

- **Frame your subject away from the edges of either lens when possible.** Since you have to mount the X3 in line with the pole for the floating camera effect, holding your camera on a pole away from you often places you near the stitch line. When possible, tilt the pole up slightly to place yourself away from the stitch line.
- **The longer the pole, the better the stitch line.** You want to keep your camera at least 10 inches from the mounting base or from your hand if you are holding the pole. If your camera is any closer than that, the stitch lines will not line up properly.
- When filming in 360, as much as you will be tempted, **don't angle the camera on the pole** just to center yourself in the frame. Due to the way the camera stitches the images, you will end up seeing a cutoff pole in your shots.
- When editing in Insta360 Studio, you can use the **Calibrate button in the Stiching Tab** to recalibrate the shot. This recalibration based on that exact shot usually produces more accurate stitching. Also, if you used any accessories, (such as Sticky Lens Guards) make sure to apply the stitching for that specific accessory.

EXPOSURE

Exposure refers to the amount of light let into your camera during recording. "Correct" exposure is usually a balance between overexposed (where there is a loss of detail in the bright, highlighted areas) and underexposed (where there is a loss of details in the darker, shadowed areas). Getting correct exposure is easiest to capture in front-lit or sometimes cloudy conditions. If you are shooting in other lighting conditions, in most situations, you may want to adjust exposure in editing software after you film.

By default, the Insta360 X3 camera sets exposure automatically. However, there are **two ways you can affect the exposure:**

1) The first way to affect exposure is to use Exposure Value Compensation in the settings. You can adjust the exposure by telling the camera to make your video or photos up to two stops brighter (+2) or darker (-2). You can instantly see the effect of your changes on the Touch Screen.

2) The second method is to **use Shutter Speed and ISO lock** to manually set your exposure, as you learned about in the video and photo settings in Step 2. The Live View on the Touch Screen is a great tool to determine the correct exposure.

AUDIO

Audio is an important element for most videos, especially scenic, lifestyle and vlog clips. The Insta360 X3 features **four microphones for full 360 "ambisonic" audio** with several options to choose from. We cover these options briefly in the settings menu, but let's go into more depth here so you can get the best audio possible from the Insta360 X3.

The Audio settings are changed in the Shortcut Menu by pulling down from the top of the screen.

The improved audio recording and controls on the Insta360 X3 let you capture amazing high-quality footage with audio to match.

There are several options for recording audio, which were covered in Step 2- Settings, but here is a quick recap:

- **Direction Focus** uses the four microphones to create an audio file that shifts focus **based on the reframed view.** All directions of audio are recorded, but as the view changes when editing in the Insta360 App or Insta360 Studio, the audio will be enhanced accordingly. This works well when you want the audio to come from the subject in view, but you will want to turn off direction focus if you plan to voiceover as you show the scenery. In Single Lens Mode, the audio will be enhanced in relation to the lens that is recording.

The **easiest way to edit your videos with Direction Focus** is to reframe the clips in Insta360 Studio or the Insta360 App and then import the edited clip into your primary video editor. Direction Focus can also be turned off with a tap in the apps if you don't need want it to be applied to the clip. Alternatively, when Direction Focus is enabled, you can edit the 4 individual audio tracks for ambisonic audio in another video editor, such as Premiere Pro.

• **Wind Noise Reduction** reduces wind noise and enhances voices for windy situations or when you are moving quickly. Wind Reduction in the X3 works really well to reduce background noise, making your voice stand out. You may hear a slight buzzing in really high wind conditions where the camera had to reduce a lot of ambient noise. Wind Noise Reduction may also reduce natural background sounds, so if you want more of a natural, full sound, choose Direction Focus or Stereo.

• **Stereo.** This mode creates a standard stereo audio file with 2 left and right audio channels.

MORE X3 AUDIO TIPS

• **Talk to your camera's mic before saying your real dialogue** to help calibrate the microphones. This will help maintain even levels throughout your dialogue.

• In windy conditions, **use Wind Noise Reduction.** This setting does a superb job of reducing wind noise while keeping your voice clear.

• If you plan to share your videos in full spherical 360 for a virtual reality experience, **record in Direction Focus mode.** The four audio tracks will give the viewer a more realistic spherical experience as they change the focus of where they look.

• If you have a specific direction of focus **in Single Lens Mode, use the microphone on the same side of the lens by using Direction Focus.** This does make a noticeable difference in audio quality, which is especially useful **for interviews, vlogging or any dialogue in general.**

• In many situations, **an external microphone** will produce even better results. **Airpods** can be connected to the camera using Bluetooth for a quick and easy external mic. Simply put the Airpods into pairing mode and press the Airpods Icon in the Shortcut Menu to connect. To connect **an external mic such as the Rode Wireless Go/Go II or VideoMicro,** use the Insta360 X3 Mic Adapter. With a cold shoe mount such as the Ulanzi PT-20 or the Insta360 Utility Frame, you can mount the mic without it being visible in the shots. The Wireless Go II with a lav mic is a great way to record wind-free, crystal-clear audio straight into the camera.

• Alternatively, you can use your phone to **record a separate audio track** to synchronize with your Insta360 video footage, but this track obviously won't have any spatial awareness. You can hold the phone close to you but out of the video frame while you video. If wind is a problem, use a lav mic (Rode Lavalier Mic is a good one) with your phone or a portable recorder. Use this technique when stereo audio is all you need.

• After you go underwater, the microphone openings tend to hold some water. **Blow them off to remove the water** from the tiny holes once you are back above water to capture the clearest audio possible.

BATTERY LIFE AND EXTERNAL POWER

The Insta360 X3 uses a removable, rechargeable 1800mAh lithium-ion battery that should give you about **80 minutes of recording time in 5.7k @ 30 frames per second.** Lower resolution and frame rate video settings will use slightly battery life. The actual recording time depends on a few factors. Battery recording times will be less in cold weather. Luckily, the battery is removeable, so when one battery runs out of juice, you can replace it with a fresh one to continue recording.

If you decide to use your phone's charger to charge the battery in-camera, the Insta360 X3 requires **a charger that outputs 5V and 3A** so check your charger before you connect it to your camera. The X3 is also compatible with external USB power, so if you want to use an external USB power source, you can run off external power for an extended amount of time. The side door doesn't have an opening for the USB cable when closed, so you will have to open the door until a pass-through door is made for the X3 with an opening for the USB cable.

In addition to charging the battery in the camera using the USB cable, you can also use the Insta360 X3 Dual Battery Charger. This charger allows you to charge the batteries out of the camera leaving your camera available for filming. The Dual Battery charger can use USB power or a wall socket and comes with one spare battery, which is very useful to have on hand.

Unlike NiCad batteries, Lithium batteries don't have a memory charge, so **recharging your battery even if it is not completely drained will not reduce the battery life.** Like all lithium-ion batteries, the battery in the X3 will eventually lose some capacity over time, so a fresh battery may be your best bet if you notice that your battery stops performing like it used to.

Follow these additional tips to preserve your battery's life:

• **Store your camera at a normal room temperature** when possible. Try to keep your battery out of extreme heat, especially when fully charged. Keeping your camera (with the battery in it) in a hot car will deteriorate your battery's capacity.

• If you need to **store a spare battery for an extended period**, use it until about 40% battery life remains and store the battery in sealed bag in the refrigerator.

There are several things you can do to preserve battery life while actively using the Insta360 X3:

• If you know you won't be filming for a while, **turn your camera off to save battery.** If you know you won't be filming for a while, it's easy to hold down the side Mode button to turn off your camera so it doesn't use any battery life. Also, setting the Auto Turn Off for a short period of time (3 min) will ensure that your camera doesn't sit idle for too long.

• **QuikCapture Mode reduces standby time** by turning your camera on and recording with one push of the button. There is a delay however while you wait for your camera to turn on, so QuikCapture doesn't work for all filming scenarios.

• The Touch Screen also uses up a good chunk of battery. If you are using the Touch Screen often and don't need the full brightness, you can reduce the screen brightness in the Shortcut menu. If you are primarily using the Touch Screen to set up your shots, **let the screen go to sleep after you have composed your angle** and tightened your camera in place. To put the Touch Screen to sleep while recording, simply push the Power Button on the side of the camera.

OVERHEATING

The Insta360 X3, like many action cameras and other high performance video cameras, will overheat if you record long enough. When recording in high resolution, such as 5.7k 360 video, the camera will get very hot. This is normal. Once it gets too hot, the camera will shut off to protect itself. This also happens with many other cameras and has to do with the fact that high resolution videos demand a lot of energy from the camera.

If your camera overheats, you just have to wait for it to cool off so the best option is to **follow these tips to reduce heating:**

• First of all, if your camera is moving with you and there is airflow, **this helps cool off the camera.** A camera sitting still on a flat surface will overheat quickly.

• Switching out batteries will extend your recording time because the **fresh battery** will start out cool. If you are recording really long clips in a dry environment, **remove the battery completely and use external power.** This will reduce heat and allow for the longest recording times.

• Recording in **Single Lens Mode or lower resolutions** requires less performance and will give you longer recording time before the camera overheats.

• **Record shorter clips when possible** to give the camera time to cool off in between recording.

• And whenever you aren't actively filming, **turn the power off** to let the camera cool down.

USING GPS

The X3 does not technically have GPS built in, but rather **works with GPS through your phone or tablet** when connected to the Insta360 App. GPS data can be added to your videos when the camera is connected to the Insta360 App only if the app stays open while recording. This data can then be used for Google Street View, but in reality, it is inconvenient to use.

The GPS setting is located in the More Settings area of the App (the three dots in the top right corner). Once you start recording, make sure the app stays open. You may need to change your phone settings to prevent the touch screen from going to sleep.

A **better option** if you are looking for live GPS data overlaid onto your footage is to use the **Insta360 GPS Action Remote.** This remote also gives you remote camera control from up to 65ft. away, which can be really convenient when your camera is mounted. After recording with the GPS Remote, import your footage into the Insata360 App to add overlay stats, such as speed, elevation, patch and more.

VLOGGING,TRAVEL, ACTION SPORTS, and REAL ESTATE TIPS

VLOGGING TIPS

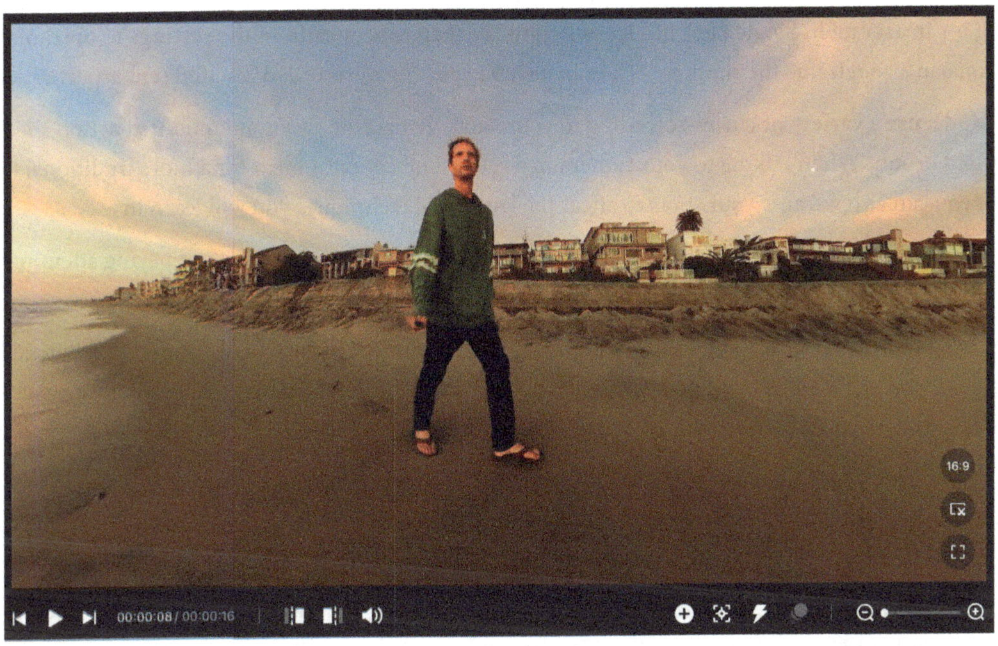

Video recording in 360 Video Mode at 5.7k @ 30 FPS. The Insta360 X3 was handheld on the Insta360 114cm Invisible Selfie Stick with a monopod base.

With the explosion in popularity of vlogs, the design of the Insta360 X3 really opens up the creative options for vloggers. The X3's unique 360 features make it **the ultimate vlogging camera,** especially if you are planning to post your traditional videos reframed to 1080p. You can also reframe to 4k when editing in Insta360 Studio on a computer, but you'll have to be a bit more cautious to maintain high quality. (We will talk about that in Step 5- Editing.)

The X3 is small, portable, and versatile. When using the X3, you can create **a wide variety of shots with a single camera** making it the perfect camera to carry with you in your pocket on your adventures.

VLOGGING VIDEO TIPS

• Both front and back lenses record equal quality video and photos. If you want to see yourself on the Touch Screen while you film, **use the lens on the same side as the Touch Screen** to compose your shots. After selecting your settings, make sure to long press the screen to hide the touch screen icons for a clear view of the scene.

• **Always be ready to be in a shot.** Smile and dress like you would want to be seen. You don't need to reframe yourself into the shots, but if you are ready and smiling, you will always have the option to include yourself.

• If you are not filming for slow motion in Single Lens Mode, filming at **24 or 30 FPS requires less light.** This will produce sharper videos in low light scenes. And with FlowState, you will get smooth, stabilized videos straight out of the camera.

• **Don't forget those other modes.** Use Video Mode for the majority of your shots, but TimeShift videos can be used to add some unique clips into the mix. TimeShift Videos can also be created from regular videos in the apps, but you will want to film with a TimeShift in mind by following the tips in TimeShift Mode in Step 2. Bullet Time videos are also a stunning addition to any vlog. For Instagram, film or reframe vertical TimeShift videos for a truly full-screen mobile experience.

• One of the biggest benefits when filming in 360 is that the invisible pole makes it appear that someone else is filming you. A **longer pole/selfie stick** such as the 270Pro or Insta360 Extended Edition Selfie Stick will give you more flexibility for reframing your shots when you edit. As you film, keep the camera around eye level or slightly higher for a better angle of you. Looking up from below the face widens the chin and can create a double chin look. Try to aim the lens towards you as much as possible.

• **Me Mode** is a great vlogging mode, when you want standard videos with the invisible pole effect. The key to filming with Me Mode is to hold the camera straight out from you from the waist. The downside of this is that you can't see your expressions on the screen as you film.

• If you don't need the floating camera effect, when filming in **Single Lens Mode** for example, tilt the camera on the pole for an easier angle to center yourself in the frame. Use the back lens to record and compose your shot on the Touch Screen.

• When recording for Instagram in Single Lens Mode, **switch to the 9:16 aspect ratio in the settings** to maximize screen space on mobile devices. You can also reframe the shots to 9:16 in Insta360's apps if you recorded at a different aspect ratio.

• In Single Lens Mode, **use a variety of lenses** according to your shots. For scenic shots and mounted action shots, use Wide for the most scenery and beauty. When talking to your camera in a selfie shot, try out Linear for a more traditional look with less fisheye. With 360, you can also reframe your shots to affect the lens using techniques you will learn in Step 5.

• Horizon leveling is another amazing feature to **keep your footage stable** even when you're not paying attention to how you're filming. The stabilization and horizon leveling capabilities make filming yourself almost foolproof.

• **Add some lighting.** Some fill light, such as a panel light or GoPro's light mod, helps to add a more professional look. When using a metal vlog case like the Insta360 Utility Frame, you can easily attach a light for better lighting on yourself or your subject. See Step 7 for more information about the best lights to use with the Insta360 X3.

When you are outdoors, **look for shady areas that are still bright.** This will give you the most even lighting on your face. Indoors, try to avoid artificial lighting. Rather, look for some **nice filtered light from windows** to light you up against a darker background. If you can find a sunlit area against a dark, shadowed background, this also adds some dramatic lighting.

• When recording in 360, since the X3 can record everything at once, you can **pull multiple angles and clips from the same shot.** Vloggers won't need to rotate the camera to show what they're looking at or talking about. Just recompose the shot when you edit and it's good to go. You can also apply MultiView in the Insta360 App for a view through both lenses individually on the same screen.

• Use **catchy transitions** between views to keep the vlog interesting. For example, if you reframe from the front to rear lens view, use a quick motion and add a swish sound effect. Also, add in some hand gestures, as if you are repositioning your camera or pointing out a landmark.

AUDIO TIPS

Since Vlogs require talking to the camera, high quality audio is vital to a successful vlog. Use the tips in the audio section of this step to record the best audio possible with the X3.

THE X3 BASICS FOR THE TRAVELER

Thanks to its compact size, crystal-clear image quality, and versatility, the Insta360 X3 makes **the perfect travel companion.** Capture your memories so you can bring them home with you, but remember these helpful X3 travel tips so you score the footage without getting bogged down.

• **Consolidate your gear.** Even if you have a full collection of mounts and accessories (because there are lots of options), **only bring what you need. Extra batteries** are always useful, along with **a charger that can be used at your destination.** Figure out if a car charger, wall charger, or maybe even a solar charger goes best with your travel plans and go with that. Pick **two or three of your most versatile mounts**- preferably at least one that can record POV angles, and your favorite pole with a monopod base- to film a variety of angles with minimal gear.

- Look at other media for an **idea of an area's highlights.** Local postcards, calendars, and artist's landscape paintings usually show an area's best attractions, so use these as inspiration. Check Instagram to see what other people are photographing in a particular area and then capture your own perspective of the sites that interest you.
- Edit your 360 videos for a **variety of perspectives.** The ability to edit multiple angles from one video means you can film more travel shots in less time than ever and create multiple photos and videos from one video clip. In Single Lens Mode, **use a combination of fields of view** to add variety. Use Linear shots for scenery and close-ups of cultural details. Mix in Wide for point of view shots and unusual angles. Mount your camera to a moped, scooter or vehicle. A variety of perspectives in your shots will keep your viewers interested longer.
- Record TimeShift Videos for exciting **travel hyperlapses.** Film a few time lapses or hyper lapses of scenic highlights for extra mood. Film from the top of a tall building, near a bridge, over a river or around city lights for a different overview. When you edit, pan around the shots for an inclusive look around your destination. Also, be sure to use **Time Lapse Mode with full 8k resolution** for longer time lapses with lots of room for creative editing!
- If you have their permission, **film people** because personalities add character to a place. Or film your travel buddies to add a subject in your shots.
- Use Direction Lock Audio to **record the sounds of a place,** such as city sounds, nature, trains, etc.

Most of all, have fun, because that will make the best videos!

X3 FOR THE SPORTS ENTHUSIAST

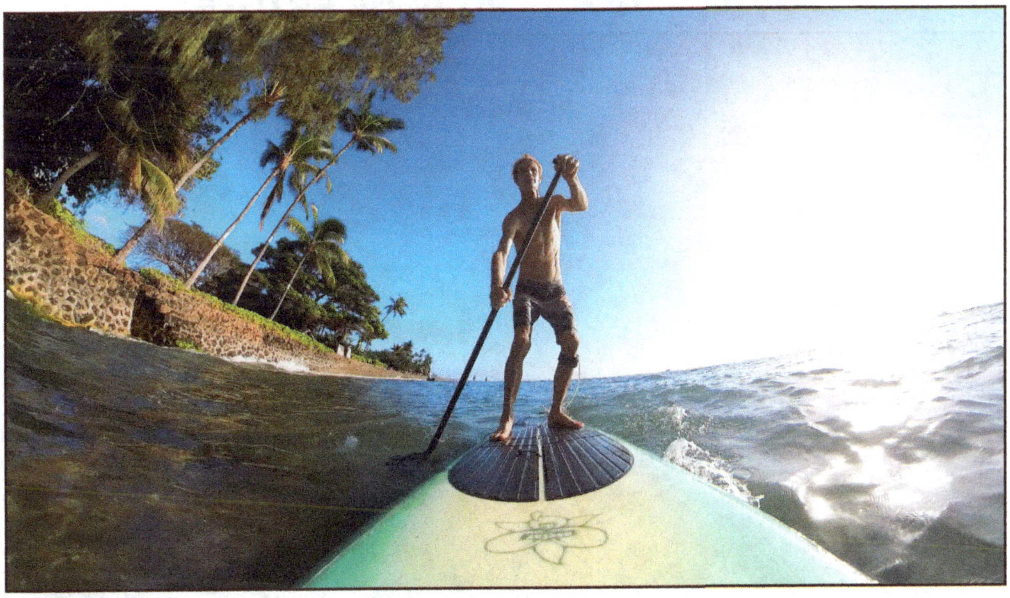

Still frame taken from 360 video recorded on the Insta360 X3 in 360 Video Mode at 5.7k @ 30 FPS for the highest quality.

The X3 has the potential to take your sports videos and photos to a new perspective, and people who watch your footage will be often ask, "How did you DO that?"

The **biggest benefit** to filming sports with the X3 is **the invisible pole.** That floating camera effect in addition to the super wide angle creates smooth follow-along footage that seems almost impossible to capture.

The **main downside** of filming with the X3 versus a traditional action camera is the **lack of super slow motion video settings in high resolution.** Most of the time, you should film in 5.7k @ 30 fps, but if you really want slow motion, 4k @ 60 fps gives you the option to slow down your 360 clips up to 2.5x. Alternatively, you can use the ultra slow motion technique you will learn in step 5 to add a few slow motion highlights to your clips.

FOLLOW THESE TIPS TO STEP UP YOUR SPORTS SHOTS:

• **Mount your camera to several different locations on your sporting equipment.** Filming different perspectives of the same activity will give you lots of creative to freedom to mix it up when you go to edit, giving your viewers a better feel of the rush you experience when you are out there in action. The **Third Person Bike Mount** is a useful mount for this type of shot, as well as an extension pole mounted to an adhesive mount. The **Suction Cup** can also be handy to have on hand. But it really depends on what sport you are into.

• **Capture POV shots.** The X3 works amazingly well for first person point of view shots that really give your viewers a sense of the action. By mounting your camera on your body, using a **Chest Mount or Bite Mount,** you can create a perspective similar to what you see. These shots take some imagination because they will look oddly distorted when you go to reframe. After reframing though, you can get the widest perspectives possible with any camera with a real sense of speed.

• **Follow the action.** Another shot you should definitely go for is a follow shot, where you mount your camera on a pole behind you, either using the **Back Bar** (a waist harness) or the **Third Person View Mount** (a backpack polecam). The combination of invisible pole and an unrealistic sense of distance from your back makes this shot one of the most awe-inspiring. It's a bit of a hassle to shoot but well worth the effort.

• **Lastly, keep your camera safe.** The X3 is not very well protected from bumps and knocks. When you mount your camera on your equipment, it's more susceptible to getting damaged. Be aware of where your camera is as you film and try to keep in out of harm's way. Use Sticky Lens Guards if you really want to protect your lenses.

USING THE X3 FOR REAL ESTATE

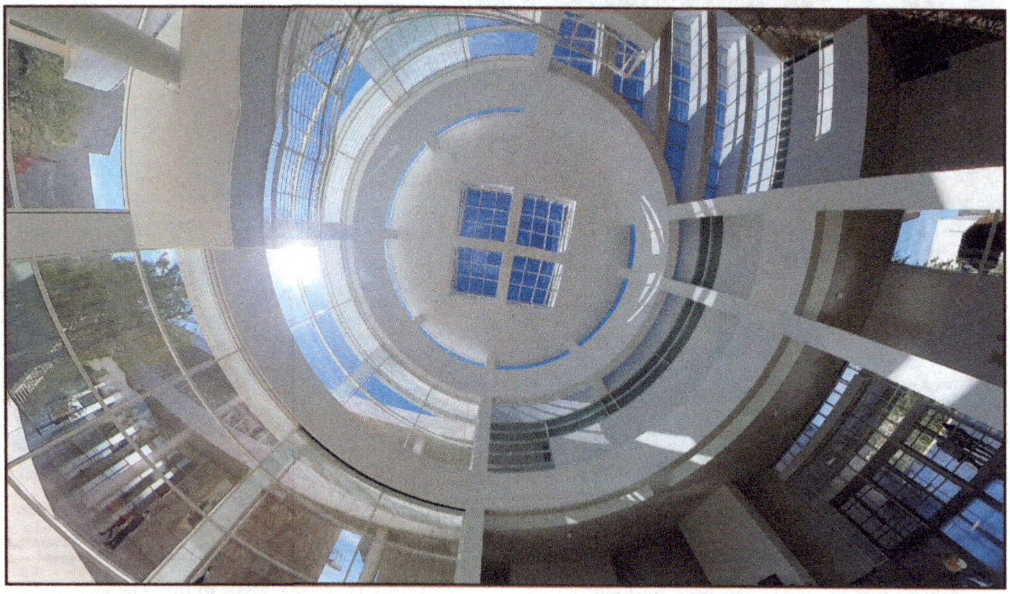

Photo taken on the Insta360 X3 in 360 Photo Mode at 100 ISO for the best quality. The camera was set down on a monopod and the self-timer was used to exit the scene before the camera took the picture.

A 360 camera like the X3 is **an essential tool for real estate marketing and virtual tours.** With 360 Virtual Tours becoming a standard component to a real estate package, real estate walkthroughs and other virtual tours offer huge potential for making money with your 360 camera.

Whether you are creating real estate walkthroughs or virtual tours of museums, art galleries, hotels, vacation rentals or other attractions, a 360 camera like the X3 gives you the tools you need to create professional-looking virtual tours.

Lots of companies are offering virtual walkthrough creation platforms, but some of the best to start with are Kuula and 3D Vista (these two are officially supported on the X3). Compatibility with Matterport will likely be available soon, if not already.

To create the photos you need for these virtual tours, all you need is your 360 camera, a monopod that extends to at least eye-level, and a little bit of knowledge.

Here are some tips to get you started creating your first virtual walkthrough:

- **Mount your camera straight up on a monopod** and extend your camera to around eye-level (5' from the ground) for the most realistic perspective of someone actually walking through a property. Set your camera on the ground when possible, but if you have to set your camera down on a surface, such as a table, still try to maintain that eye-level view. Try to avoid any reflections that will show your camera to reduce the amount of editing you will need to do later.

- Try to set your monopod on **an easy area to clone** out the monopod feet. For editing, a solid color floor or table is much easier than a complicated tile pattern, for example. We will show you how to clone in Step 5- photo editing.

- **Set the stage.** Before you take any photos, look around to make sure the property is clean, empty of people (so you don't have to sign model releases) and as beautiful as possible. It's not technically your responsibility to stage the property, but your clients will be happier, and you will look more professional if you deliver imagery of a presentable space. No one is going to be happy if there is a towel laying on the ground, for example. Also, turn on all of the interior lights before taking photos.

- **Use 360 Photo Mode in 72MP** to take the high quality photos needed for the virtual walkthrough. Take photos with the camera facing the same direction when possible. Use a low ISO of 200. With the constant mix of lighting sources, setting a manual white balance setting can be tricky. For the best white balance settings, take a picture first using a gray card. Place this card near an area that you want to make sure the White Balance is accurate, near a wall for example. Then, take the same photo again without you in it. You can match the white balance from the first image to the second image using Photoshop or Lightroom. Set the self-timer to 10 seconds and disappear from the scene. Or use the WiFi Remote or Insta360 App (if the distance is not too far).

- **Plan out your walkthrough.** You want to give viewers a smooth flow as they move through a potential purchase property. Place the camera near the center of each room so they can view all of the areas from a relatively equal distance. Also, always plan one shot ahead. Try to take one photo about every 10 feet for apartments and one photo every 20 feet for larger spaces. As you take a photo, you want to make sure you can see the next camera location. That way, when someone is moving through the tour, he or she can see the next point and easily move through the property.

- Once you have a complete tour of high quality images, **use one of the walkthrough apps listed above** to deliver the goods and make your clients happy.

Now that you have the knowledge to capture your footage, get out there and start filming! In Step 5, you will learn how to edit your photos, videos and time lapse clips!

STEP FIVE

CREATION

Now It's Time To Put It All Together

So, now you've got the footage! Editing with the X3 is one of the most exciting steps of the process. The creative possibilities are unlimited! This is where you are able to take your raw footage and translate it back into an experience both to remember and for others to enjoy!

In this step, we will first show you a variety of editing techniques for editing your Insta360 X3 videos. Then, we will show you some of the essential techniques for editing Insta360 X3 photos. Follow the steps below to create your own masterful 360 content!

BEFORE YOU START EDITING

TRANSFERRING FOOTAGE

If you are using the Insta360 mobile app, you can edit files that are on the camera's microSD card or files that are stored in the Album library. On a computer, you can also edit files that are on your camera or files that have been transferred to your computer or another storage drive.

Saving Files To A Phone/Tablet

The Insta360 App is the easiest tool to use for transferring your footage to a phone or tablet. **Follow the steps below to transfer your X3 footage to your phone or tablet:**

Step 1: Turn the camera on. Open the Insta360 app and Tap the Album Icon at the bottom of the App. On the Top Right corner, press the Camera/WiFi icon to connect to the camera. Follow Insta360 X3's connect dialogue.

Step 2: Select View Media to connect.

Step 3: Tap on the photo or video thumbnail to preview a low-resolution version. The images may appear to be slightly blurry because of the low resolution .LRV file used for the preview, but don't worry, the final edits will be full quality. You can edit the files without saving them to the phone first.

Step 4 (Optional): For any files you want to save to your phone/tablet, Tap the Download icon on the far right of the editing icons. This saves the high-res file to the Insta360 App's media library on your phone or tablet. You can also edit the files while they are still on the camera's microSD card and export the hi-res edits to your phone/tablet.

TIP: If you want to edit a photo or video outside of the Insta360 App, open the file in the media library. Then tap the Export icon to save the file to your phone's photo library.

TIP: **Quick Edits.** For an even quicker editing experience, use the **X3 Quick Card Reader.** This plugs directly into your phone, eliminating the need for WiFi and speeding up the transfer/editing process. When using the Quick Reader, in the Shortcut Menu>Settings/USB Mode, select Quick Reader.

TIP: If you have already downloaded the footage off of the microSD card onto a computer and still want to edit the files on a mobile device, you can text or email them to yourself from your computer and save them to your device's photo library. However, you won't be able to edit them in the Insta360 App.

Saving Your Files To a Computer

If you plan on editing your photos or videos on a computer, or just want to back up the files to a computer, **follow the steps below to transfer your X3 footage to your computer.**

Step 1: First, in the Shortcut Menu>Settings>USB Mode, select U-Disk Mode. Connect your camera to the computer using the USB cable.

Note: You can alternatively remove the microSD card, insert it into an SD card adapter, and insert it into a SD slot on your computer.

Step 2: The microSD card icon will show on your desktop screen. It should be called Untitled. **Double click** to see the contents. **Double click again on the DCIM folder.** The files are located inside the folder/folders located there. Drag the files from each folder over to a new folder on your computer.

Note: The .LRV (low resolution video) files are used with the Insta360 App so you don't need to keep them on the computer. Videos end in .insv for 360 files or .mp4 for Single Lens Mode files. Each 360 video file has two files, so make sure you transfer both of these. Photos are .insp and .dng (for RAW files).

TIP: **Organizing Your Files.** If you don't already have a format for organizing your media, you may want to set up folders by date. Make a year folder with month subfolders. Within those you can add the day if you want. Since the .360 files are hard to preview as a thumbnail, you may also want to add a subject name on the folder.

Step 3: View your transferred pictures and videos. The pictures and videos you just transferred are now copied onto your computer.

Step 4: Now you are ready to start editing. (You can also edit the files that are stored on the camera's microSD card without transferring them, but the camera will need to be connected if you want to go back to your edits later.)

STORAGE

Your X3 files are going to require a lot of storage space. You will probably want to **dedicate an external hard drive to your X3 files** (or at least a big block of memory). How much storage you need really depends on how often you use your camera, but a portable 5TB hard drive is a good starting point. Go bigger if you can. An external hard drive is the easiest way to keep your full resolution files on hand. If you do most of your editing on a device, you will need storage available on your device or on an external drive. And remember to always back up your files to cloud storage or another external hard drive because you don't want to lose all of your hard-earned memories.

DELETING FILES

There are several methods to use to delete the files from the microSD card after you have transferred them.

On the Camera

Using the Touch Screen:

1. Swipe Right to view your files and select the Grid Icon.
2. Select files to delete one-by-one or select all files by tapping the check mark.
3. Tap the Trash Can Icon to delete selected files.

To format the microSD card:

1. After you have copied the files from the camera to your computer, swipe down on the Touch Screen to bring up the Shortcut Menu. Then tap the Settings Icon
2. Scroll to the bottom of the Settings until you see the SD Card option. Under SD Card tap Format.
3. The screen will ask if you want to Delete all files. Tap the check mark to confirm.

Use the Insta360 App

You can also copy and delete files using your smartphone or tablet.

To erase individual files:

1. With your camera connected to the Insta360 App, view your Album in the App.
2. When you select a video or photo file, tap the Delete icon on the right side of the editing menu.

To erase all of the photos and videos using the Insta360 App:

1. With your camera connected to the Insta360 App, in the Album, tap the check mark at the top right corner.
2. Select all of the files and tap the Trash Can icon.

On a computer

When your camera is plugged into the computer with a USB cable, you can delete files from the microSD card by moving them to the trash can.

VIDEO EDITING

In the video editing section, you will learn exactly how to edit your X3 video clips into a short video using FREE apps when possible. Whether you want to post a single clip to Instagram or create a fully-edited video, editing (also known as "post-production" in the video world) is where you **take the results of your creative filming and put together the pieces to share your story.**

Luckily, most of the tools you need for producing professional-level content are available for free: you just need to know where to look and we will show you. There has never been a better time to get started with your 360 video editing!

When editing your 360 videos, there are **two formats for outputting your videos, either as a 360 spherical video, or as a "reframed" standard video,** such as horizontal 16:9, vertical 9:16, or square 1:1.

The output you choose has a huge impact on how you edit your videos. Reframed videos offer more editing creativity and are the most popular for standard viewing. 360 spherical videos give your viewers the power to look around the scene to choose their view for a full 360 experience. We will cover both types of editing in this step.

First, we're going to show you how to edit your 360 videos on a mobile device, such as your phone or tablet. We will be using the Insta360 App since the 360 editing tools on the app were designed to work with the Insta360 X3.

Once you learn the basics, editing 360 videos using the Insta360 App on a mobile device is quick and relatively easy. Mobile editing, however, does not produce as high quality videos as computer editing. (The Insta360 App limits reframed exports to

1080p, while in Insta360 Studio, you are able to export up to 4k.) If you're editing videos for Instagram, mobile editing is fine, but for YouTube or professional level content, you will want to use a computer to edit your videos.

Then, **we will move over to computer editing** to show you the next step in high quality 360 postproduction. Many of the techniques you learn about for mobile editing transfer directly over to desktop editing, so we can begin with the basic concepts using mobile editing.

Editing on a phone or tablet is totally optional. Everything except for the Insta360 trick shots (Shot Labs) can be done on a computer as well. Even if you plan to edit solely on a computer, **please read the mobile editing section to understand some of the basic concepts that are so vital to editing 360 content.**

MOBILE EDITING

Follow these step-by-step instructions to edit your 360 videos using the Insta360 App:

 # 1- Open Your File

Now it's time to start editing. Let's open the file you want to work with.

1.1 In the Media Gallery on the Insta360 app, **select a 360 video file** to begin with from the Media Gallery or from your camera's microSD card.

1.2 Tap the thumbnail to open the file.

#2- Understand The Editing Options

The Insta360 app appears to be very simple but actually offers more edit options than meet the eye. Let's take a look at the editing icons to see what is possible when editing using the Insta360 app.

At the top of the screen, the Insta360 App offers three main editing methods: Auto, Snap, and Edit.

2.1 Option 1 Auto- Auto uses the App's AI to search through your clip and automatically pull out and frame any clips it thinks you would want. Simply **press the AI Analysis Icon and see what happens.** Then tap on the extracted clips to preview them. Whether the app actually pulls out the clips you want or the perspective you are looking for is hit or miss. But, if you are looking for some quick, easy edits to share, it's worth a try to see what the App produces.

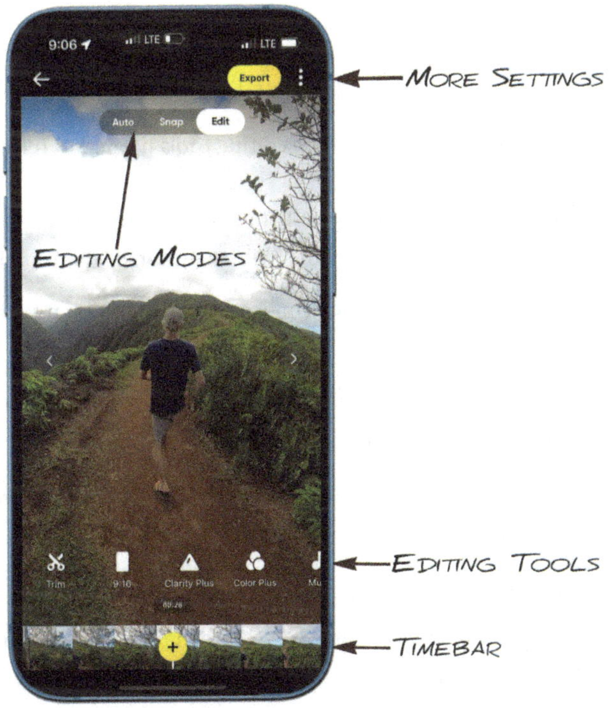

2.2 Option 2 Snap- Snap enables you to reframe your video based on what you see on the screen. To reframe videos in Snap Mode, press the record button and either physically move your phone around and/or drag your fingers on the screen to move the image. The movements and adjustments you make are recorded and will be saved as a reframed video. This method works okay for slower paced videos, but it doesn't match the control you have using the full Edit mode.

2.3 Option 3 Edit- In Edit Mode, all of the editing tools you need to **produce professional-looking edits** are there at your fingertips. Since this editing mode gives you the most creative control and teaches you the necessary tools for editing 360 footage, let's look deeper into editing using this mode.

TIP: More Settings. Under More Settings at the top right of the app, you can turn on or off some of the automatic features. Let's take a quick look at these since they have a huge impact on your videos:

Flow State Stabilization- FlowState Stabilization keeps the horizon level and removes any shakiness caused by camera movement. Only turn this off if you want a more raw video or if the video has jittery stabilization effect caused by low light scenes.

Direction Lock- Direction Lock matches the viewing direction with the lens you are looking through.

Direction Focus- If you recorded audio using the Direction Focus setting, this can be turned off here, which is useful if you are narrating from behind the lens, for example.

Identify Tracking Targets- Places crosshairs on objects that can be tracked.

AquaVision 2.0- Insta360's automatic underwater color grading can be enabled for underwater footage.

Chromatic Calibration- Creates a better balance of exposure between the two lenses. Under Accessories, you can adjust the stitching for accessories such as the Sticky Lens Guards and the Dive Case.

Accessories/Stitching- Affects Stitching for Sticky Lens Guards and Dive Case.

#3- How to Navigate your 360 Videos

The first thing you will want to understand is how to adjust the composition so you can plan a rough vision of what you will edit. On a phone or tablet, there are three ways to adjust the composition, which are all pretty intuitive.

3.1 To **Pan and Tilt** around the scene, your fingers on the screen to move the image.

3.2 To **Zoom In and Out**, pinch your fingers together or apart just like you would to zoom in on one of your phone's photos.

3.3 In the primary view, the **Horizon is locked.** We will show you how to rotate the horizon when we talk about using keyframes.

#4- Using the Editing Tools

This toolbox gives you all of the editing tools needed to reframe your 360 videos. Let's take a look at how they work:

4.1 **Trim**- Use this icon to trim a section out of the full video. This is especially useful to trim off the beginning/end of the clip if you physically pressed the Shutter Button or the Vibration made a sound you would like to delete.

To trim your videos, tap the Trim icon and use the Trim tool (not Jump Cut). Slide the timeline to scrub to the moment you want your clip to begin. Then tap the trim icon in the timeline.

Slide the timeline to the moment you want to end the clip. Press the trim icon again. Note: you don't you don't need to choose your angle or perspective yet.

Press the check mark to return to the editing screen. If you want to undo your trim or modify it, tap the trim icon again to modify your selection.

TIP: Jump Cuts. On the Trim Screen, Jump Cut is used to cut multiple segments from the same clip. Then, when you reframe each segment and add a Keyframe, the framing will jump from one view to another without transitioning (hence the name "Jump Cut.")

4.2 **Aspect Ratio-** (9:16 is automatically displayed for 360 videos). Use the Aspect Ratio icon to select your desired output aspect ratio for a reframed video or change the Aspect Ratio of a video shot using a Single Lens Mode. You learned about Aspect Ratios in Step 2. Select either 9:16, 16:9, 1:1, or 2.35:1. After you reframe, you can come back and change the Aspect Ratio if you want multiple formats of the same video.

4.3 **Clarity Plus**- Tap this Icon to add extra clarity to your video. Clarity is similar to Sharpness.

4.4 **Color Plus**- Color Plus boosts the vibrance of the video clip. Once you tap Color Plus to apply it, adjust the slider to decide how much color you want added. Color Plus cannot be used simultaneously with Clarity Plus.

4.5 **Music**- Add Music for personal use from Insta360's music library. Typically, you will want to add music later to multiple clips.

4.6 **Speed**- Adjust the playback speed of the video. The Speed tool can be used to slow down or speed up a clip. When slowing down clips, **only clips recorded at a higher frame rate should be slowed down.** When you slow down a clip recorded at 30 FPS, the video will play back choppy. If you recorded at 60 FPS, use 1/2 speed. For 100 FPS, you can use 1/4 speed and still have smooth playback.

Any clips can be sped up for a TimeShift effect. Choose between 2x up to 64x the original speed, depending on the clip you are using. The Motion Blur button will add in Motion Blur for your TimeShift videos to give it a speedier feel.

4.7 **MultiView**- MultiView is a fun tool for vlogging clips because it shows your viewers two perspectives from one clip. First select a portion of the clip to display selfie view (which will be framed automatically). Then select the MultiView style. Small window displays a small window of the selfie view in the corner. Split Screen splits the screen 50/50. Car MultiView is specifically for car shots, where you want a view looking in the car and looking forward out the window. Since Car MultiView cannot be edited further, you should film for this shot in mind when you set up your camera on a dashboard.

4.8 **Freeze Frame**- Freeze Frame is a great tool for highlighting a key moment in your video, especially if you recorded at a regular speed frame rate. This tool allows you to freeze a moment of time in the middle of a video clip. For 3, 5, or 10 seconds. You can then reframe, or move around the still frame, before continuing the video. The video's audio will be muted for the freeze frame since it is only a single frame.

4.9 **Snapshot**- Use SnapShot to **pull a still frame from video** to use as a photo. If you want to pull a reframed photo from video, first reframe the shot, then tap the SnapShot tool. You can choose to save a current image or 360 image (any reframing effects won't be applied) to your phone's photo library.

4.10 **Face Filter**- Face Filter offers tools for adjusting facial tones and features, but really only can be applied to shots that are relatively close to the subject.

4.11 **Filter**- This setting offers a batch of preset color grading that makes it really easy to edit the look of your videos. If you would prefer automatically color-graded videos, scroll through the options to preview the different looks. Use the slider to adjust the strength of the filter on your shot.

4.12 **Adjust**- The Adjust tool gives you all of the standard adjustments needed to manually color grade your videos, including Exposure (brightens or darkens), Color Temperature, Contrast, Saturation, Highlight, Shadow, Tone, and Sharpness.

4.13 **Mark**- Add a marker to key moments on your video. This will place a small marker on the timeline for you to reference or for the app's AI to use for autoedits.

4.14 **Reset Edits**- Reset all of the edits you made to the clip.

4.15 **Delete**- Delete the current file.

4.16 **Download**- Download either the trimmed clip or full video to save to the Insta360 Album.

TIP: Exporting 360 files from the Insta360 app. If you want to export your video as a 360 spherical video, you can use the Filter or Adjust tools above to adjust the video's appearance. You can also trim the video using the trim tool to cut out unwanted parts. The other tools apply to reframed videos.

Because you're exporting the full 360 sphere, any compositional adjustments you make will not affect the final video. If you export your 360 videos as a spherical video from the Insta360 App, you can then edit this file in another editing app. However, exporting a 360 spherical video from the Insta360 App will result in a smaller video resolution than if you convert your files using a computer app, like you will learn to do next. The exported video file resolution is 3840x1920 from the Insta360 App vs. 5760x2880 on a desktop app.

#5- Reframe Your 360 Video

Now that you understand the Insta360 App editing tools, let's talk about **the BIGGEST editing tool** when it comes to reframing 360 videos. The Keyframe Icon is the way to really begin reframing your 360 videos to a traditional video. This is where the creativity can start to flow freely.

If you are new to video editing, keyframes are probably a new concept. Keyframes are essential to reframing your 360 content. A keyframe is used to save the composition (zoom, pan/tilt, rotation) at a specific moment in the video. As the video plays from one keyframe to the next, the composition will transition to the next keyframe's settings.

For example, if one keyframe is set to a zoomed in view and the second is set to a tiny planet view, as the video plays, it will slowly transition to the tiny planet view. There are several options for adjusting the transition which we will cover next. **Follow the steps below to begin reframing your video.**

5.1 Keyframes- On the timebar, scrub (drag the video strip sideways) to the moment you want to add the first keyframe.

5.2 + Tap the **keyframe (+) icon** and a new set of options will appear directly above the timebar. Let's take a quick look at these:

The top row offers **preset looks,** such as Tiny Planet and a variety of lenses. **Viewfinder** lets you frame the shot on the screen as the video progresses. This can be performed for even just a short portion of the clip. Select **Deep Track** when you want the view to automatically follow a subject. Simply drag a box over the subject and the App will start tracking the subject.

TIP: Deep Track. You can also long press on a subject at any time to pick the subject and start tracking.

5.3 Compose your shot using the lens presets or manually adjust the composition on the Touch Screen. If you need to rotate the horizon, you can do so here using the Rotate Icon. Once you are happy with the composition, tap the Update Keyframe icon on the timeline to save the keyframe.

5.4 If you make any more adjustments to the composition at that keyframe after saving it, be sure to tap the **Update Keyframe icon to save the changes.** Otherwise, the changes won't be applied.

5.5 To **add another keyframe,** scrub on the timebar to another point in the video. Readjust the composition and tap the keyframe icon (the yellow circle with the + symbol) near the playback time.

5.6 You've just **created a transition between one keyframe and another.** Now, play your video to preview the transition between the two keyframes. **That is the basis of reframing 360 videos.**

A Few More Reframing Tips:

- To prevent the shot from transitioning from one keyframe to the next, add a second keyframe at the end of the section you want to stay the same. Don't make any changes to this keyframe and the framing will remain the same between the two keyframes.
- The framing before the first keyframe or after the last keyframe will match the closest keyframe.
- × To delete a keyframe, scrub to the keyframe on the timeline. Tap the yellow x to delete that keyframe.

TIP: Three of the Best 360 Reframes

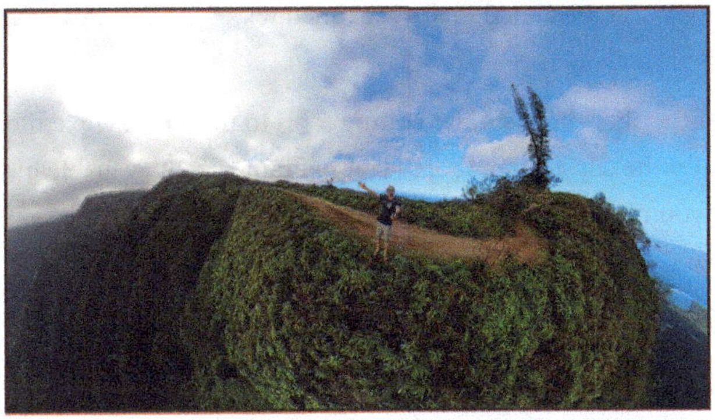

Ultra Wide-Angle: These are some of the coolest shots you can create with the X3 because they are beyond what you can do with any single-lens wide angle camera. They are great from action shots with movement and scenic shots. First, zoom out far. Then pan to find your subject. The tilt to place the horizon near the center of the frame. The straight horizon gives the shot the feeling that it's not so distorted, but the amount of scenery and distance you can create between the camera and the subject is amazing!

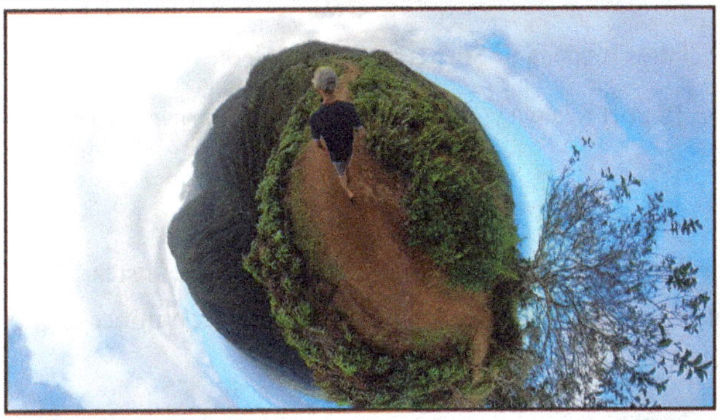

Tiny Planet: Tiny Planet shots are easy to reframe by zooming all the way out. Then drag up from the bottom to distort the horizon into a tiny planet. Or drag down from the top to create a Rabbit Hole shot. You can also just use the Tiny Planet preset and adjust the framing as desired.

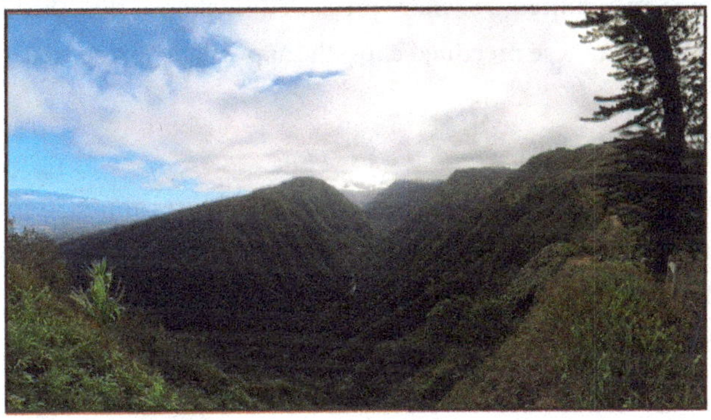

Traditional Video: Framing your 360 shot into a traditional-looking video works wonderfully for tricking the eyes. With all of the creative transitions and multiple angles you can pull from one clip, viewers will wonder how you did it. Create the appearance of a traditional video by using the Linear or Narrow lenses. Keep the horizon flat and level near the middle of the frame. You don't want to zoom in too far or reframe the video beyond 1080p when using this technique because you will start to see a reduction in quality.

#6- Exporting Your Finished Video

When you have finished editing, press the Export button. This brings up the Export dialog. Reframed videos are limited to 1080p resolution and 360 videos are limited to 4k (if you select the YouTube icon). Use the highest bit rate the app allows.

You can edit together multiple "reframed" clips using the Stories editor in the Insta360 App or use another editing app to combine multiple clips.

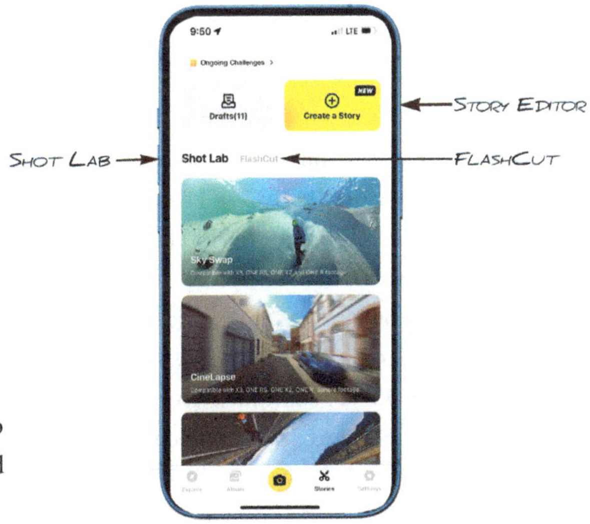

The Insta360 App Stories, FlashCut, and Shot Lab

The Insta360 App Stories feature gives you more editing options. If you tap on the Stories icon at the bottom of the App, you can access the Story Editor, Shot Lab, and Flash Cut. You will love these tools!

The Story Editor

The **Story Editor** provides you with the tools to combine multiple video clips and photos into a video edit. You can also edit individual clips, add music, titles, and stickers.

To begin editing a multiple clip "Story" video on the App, **Tap "Create a Story"** from the home page of the Insta360 App. You can then select multiple media clips from the App or your camera's media.

All types of media, including 360 videos, reframed videos, and photos, can be added to a Story timeline. (Make sure All is selected in the top left corner if you want to add photos and videos to the Story.)

Once you select your media, the Story Editor will open up. This editor offers very similar editing tools to the single clip editor that you just learned about. Tap on a video clip in the timeline to open up the clip editing tools. (Freeze Frame is not available in the Story Editor). You can then edit individual clips within the app. If you need to add more clips to the timeline, tap on the "+' icon at the end of the timeline.

Additionally, the Story Editor has a wide variety of transitions, which add a more polished look to your video. To add transitions between clips, simply tap on the small icon between two clips in the timeline, and select the transition you want to use. You can preview the transition and adjust the duration of the transition.

The Story Editor also has three additional tracks for adding Music, Titles (text), and Stickers to your video.

Once you have finished editing your Story in the Insta360 App, Tap Export to save your video. You can then save it to your photo library or share it through Social Media outlets.

FlashCut

The second tool in the Stories Editor is called FlashCut. **FlashCut offers predesigned themes for your Stories.** Instead of editing your videos manually, this allows you to plug in your own clips into a template you like. Each theme has a set number of clips and Aspect Ratios that can be used, which is shown under the title of the theme.

FlashCut videos **don't offer any extra editing tools once the video is auto-edited.** For more precise edits, choose clips that are shorter and contain the moments you want to show. Then, when you select these preedited clips, the app will be forced to use your selected moments.

ShotLab

The third tool is called Shot Lab, and **this is where the Insta360 App's magic AI really shines.** Shot Lab features about 30 special effect shots, most of which could only be created with a 360 camera. These effects would be difficult to manually edit, so if you want these types of shots, it's best to utilize the app's AI editing. Also, these are not available in Insta360 Studio. The Shot Lab exports are limited to 1080p, but they are an amazing tool for some fun trick shots.

To use Shot Labs, **tap on a Shot Lab look you want to preview the effect.** The Shot Lab gives you instructions on how to film for that particular shot. Once you have the clip recorded, select "Use This Theme," and select your shot. Follow the on-screen instructions and watch the App produce some really creative results.

TIP: More (single lens) video editing apps. There are a few other mobile video editing apps you may find useful. The Insta360 App can be used for all of your edits, but if you are looking for more, these apps offer some extra tools for editing your reframed videos (after you have reframed them in the Insta360 App.)

- **CapCut** (iOS and Android) is one of the best free video editing apps available. Made by TikTok, this app offers tons of creative transitions and fun tools for creating multi-clip edits.
- **Adobe Premiere Clip** (iOS and Android) is another full feature video editor with an option to auto generate a video like the Insta360 App or manually edit your videos. One of the best aspects of Adobe Clip is the large selection of filters to choose from.
- **VideoLeap** (iOS) is free with more tools available in the paid version. With its variety of tools and ease of use, VideoLeap is one of the best mobile video editors available.
- **Chromic** (iOS) has some nice color grading tools.
- **Post Edit** (iOS and Android) gives you the tools to edit videos as well as overlay a drawing on the video or add sticker graphics.

Check out these tips to make your mobile videos more fun to watch:

• **Combine X3 clips with other videos** taken on your phone or other camera. If you choose to edit with a third party app, any videos or photos in your Camera Roll can be used in the video.

• **Add titles.** Give your videos some character with titles or subtitles.

• With mobile videos, **longer is not always better.** You can often make the point in a short 60 second clip, which is the maximum length of an Instagram video.

• Choose the **music for the mood.** Make sure to avoid using copyrighted music without permission. See the next tip.

> **TIP: Choosing Music.** When putting together a video for YouTube, you must have permission to use music. A few sites give you permission to use their songs for free (they usually ask that you give them song credits in exchange for the free music). Try out http://incompetech.com, http://www.melodyloops.com, and http://freemusicarchive.org which offer some free songs to use for your videos if you plan to upload them to YouTube. For free sound effects, try https://www.freesound.org or http://soundbible.com/royalty-free-sounds-1.html
>
> YouTube also has a selection of free songs under Creator Studio>Create>Audio Library. Or you can buy songs at stock sites like Envato Elements or iStock.com (they also have a free download every week if you sign up for an account).

DESKTOP VIDEO EDITING

Now that you've learned how to edit your 360 videos using the Insta360 App on a phone/tablet, let's show you how to edit X3 videos using a computer. The benefit of editing on a computer is that you have more control over the export size. Insta360 Studio allows you to **export reframed videos in 4k resolution,** which is great for YouTube when you want high quality content. Also, the larger screen size on a computer makes it easier to accurately preview your videos.

There are lots of options when it comes to video editing. **The goal of this section is to help you understand the various techniques used to edit video footage filmed on your Insta360 X3.**

We will break down the video editing into three steps, Pre-Editing, Reframing, and Exporting 360 Spherical Video.

For the following video-editing tutorials, select two or three clips you would like to work with. You can use the same clips for all three tutorials so you can see the difference in editing techniques.

Here are some tips to consider when you first start to think about how you are going to edit a video:

• **Have a vision** for your finished product. If you plan to put together a longer video, create a storyboard or timeline to plan ahead and look for the pieces you need.

• **Bookmark** your edit worthy moments. When organizing clips to use for a video, write down the time marker of your highlights for the clips you decide to use to make it easy to find these moments later. Alternatively, you can Mark moments on the App as you review your footage. This is particularly useful when you start to edit longer videos.

• **Erase files as you go.** If you look through a video clip and see that there is nothing usable, erase it as you go. You will have lots of files to sort through, so the more you can thin out your library, the easier it will be to find your usable clips. Staying organized will make your task easier when you go to edit a video.

EDITING IN INSTA360 STUDIO

Straight out of the camera, the 360 videos files from Insta360 X3 are **not compatible with other video editing apps.** Before you import your Insta360 X3 files into a video editor, you **need to convert your .360 video files** into an equirectangular format that is compatible with a video editor.

Insta360 makes **Insta360 Studio**, which is a free download for owners of an Insta360 camera. Insta360 Studio can used to stitch 360 files, convert 360 files to an equirectangular format, and also for reframing.

Insta360 Studio can be downloaded from the Downloads page on Insta360's website (for Mac/PC). If you plan to edit on a computer, go download Insta360 Studio for your computer's operating system.

Insta360 Studio

You can either choose to use Insta360 Studio as your final reframing tool or use the app to convert 360 files to an equirectangular file for another editing app.

Insta360 Studio gives you the tools to reframe your 360 content and to export 360 videos almost exactly like you learned using the Insta360 App, except on a computer. You can also use Insta360 Studio simply to export 360 videos as equirectangular 360 spherical files for further editing in another editing app. You will need to use a video editor to combine multiple clips and add text, music, etc. For reframed videos, once the videos are reframed, they can be edited just like any traditional video.

> **TIP:** Video Files. Insta360 X3 creates two video files for each 360 video using an .insv format to create files that are small enough to process inside Insta360 X3. However, these files are not directly editable in other editing apps, hence the need for conversion.

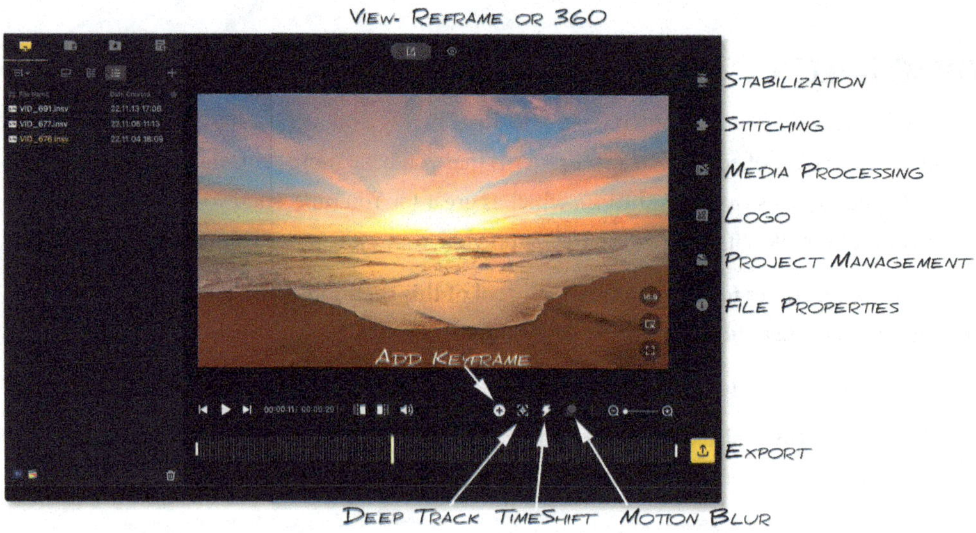

The Insta360 Studio Editing Screen

Import Your Files

To prepare or edit your 360 files to an equirectangular format, follow the steps below:

1. Open Insta360 Studio (MAC/PC).

2. Import the Video File by dragging and dropping one of the .insv video files into the column on the left of Studio.

3. Preediting Tools. There are a variety of tools in the column on the right you should consider whether you plan to export your video as a 360 file or reframe the video in Studio. Many of these were covered in "More Tools," but let's look at the additional tools in Studio. (The options will vary slightly depending on your original settings when you recorded the media.)

The tools from top to bottom are:

Stabilization- Stabilization offers the same tools as the App.

Stitching- In addition to the accessories from the App, there is an option for "Custom," where you can use a slider to customize the stitching. Under Stitching Optimization, you also have the option for Dynamic Stitching. Dynamic Stitching generates a customized stitch for each frame and works using a different algorithm that Optical Flow Stitching. Typically, Optical Flow Stitching is the best option, but try out Dynamic Stitching if you are experiencing any stitching errors. The last extra feature in Studio is Stitching Calibration. Press this button when you can see an obvious stitch line and the app will recalibrate the shot for that scene.

Media Processing- Color Plus, Clarity Plus and Aqua Vision 2.0 are located here. True Audio gives you the option for Voice Focus or general Noise Reduction. Direction Focus can also be turned off here.

Project Management- Allows you to organize and start new projects.

Info- Displays the recording data, including frame rate.

4. Next, use the **Trim tool** to trim out any unwanted portions of the video. This reduces the export time and file size by creating shorter clips.

5. Add TimeShift. If you are exporting a 360 video without reframing the only other tool you may want to use is TimeShift, which is located below the preview window. If you want to add a TimeShift, click the TimeShift Button. Select the portion of the clip to speed up and decide on the speed. Also, if you want Motion Blur added to the TimeShift, which creates a more realistic effect, click the Motion Blur Icon.

6. Export 360 Videos. Next, if you are exporting to a 360 format without reframing, Click the Export Icon. In the export dialogue, select "Export 360 Video," and the location you want to save the file. The full resolution 5.7k video is 5760 x 2880. If you plan to edit the clip in another editor (such as Premiere Pro, Davinci Resolve, or Final Cut Pro), use ProRes for the highest quality export. Use H.264 with a bit rate of around 110 if you plan to post the video without further editing.

• If you want to learn how to edit a full video in 360 for VR, export and **save two clips to a folder** to use for the 360 Spherical Editing tutorial video tutorial.

If you are reframing, let's move onto adding some keyframes in Studio.

7. Reframing in Insta360 Studio. The editing features on Insta360 Studio are **almost identical to the Insta360 App**, so now that you learned how to edit using the Insta360 App, Insta360 Studio will be easy.

Since the editing techniques in Insta360 Studio mimic the Insta360 App, you can apply the mobile editing techniques you previously learned to Insta360 Studio for reframing your 360 videos.

Readjusting the framing must be done using the mouse since computers don't usually have a Touch Screen. To navigate your 360 videos on a computer, drag the mouse sideways or up and down. To Zoom in and out, roll the wheel or slide your fingers up and down on the mouse:

8. First, select the Aspect Ratio- 1:1, 9:16, 16:9, 4:3, or 2.35:1.

9. Next, make adjustments to your composition and **add keyframes using the Keyframe icon** below the preview window. Once you add a Keyframe, a dialog box will pop up with preset lens looks as well as custom parameters. The custom parameters include the **Pan Angle** (side to side), **Tilt Angle** (Up and Down), **Rotate Angle** (which can be rotated multiple times for a spinning effect), **FOV** (zoom in or out), and **Distortion Control** (affects the lens distortion). Hover over the curve to see the name of each.

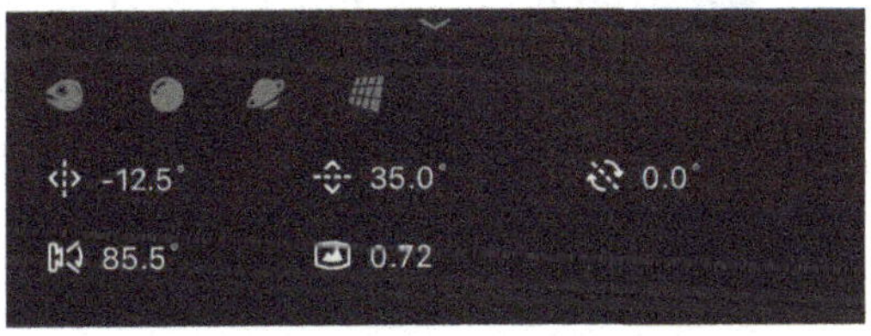

TIP: Reframing for 4k. If you plan to export your reframed clip to 4k (3840 x 2160), don't zoom in too far. Zooming in too far on 5.7k 360 footage will result in a low quality 4k export.

10. Transitions. Once you have made two keyframes, you can **adjust the transition** between the keyframes by clicking on the yellow line between them. Instead of a steady transition from one keyframe to the next (called a Smooth Dissolve), you can adjust the curve of the transition. Try them out for yourself, but Fade-In Fade-Out is one of the smoothest ones to use.

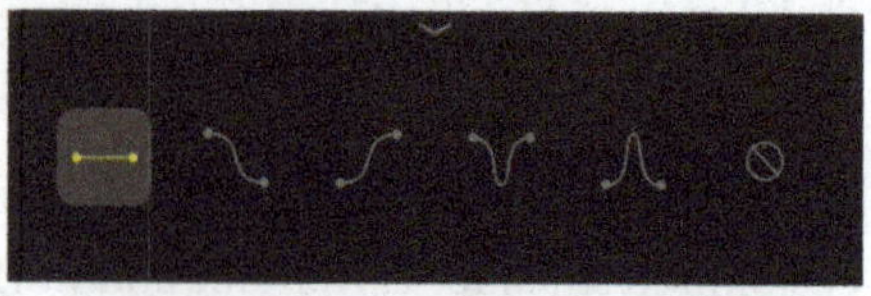

11. Export. After making adjustments to your video, click the Export icon to bring up the Export dialogue. For 1080 exports, select 1920 x 1080 resolution. If you want to export to 4k and you didn't zoom in too far when reframing (because this can affect the quality), use 3840 x 2160 resolution.

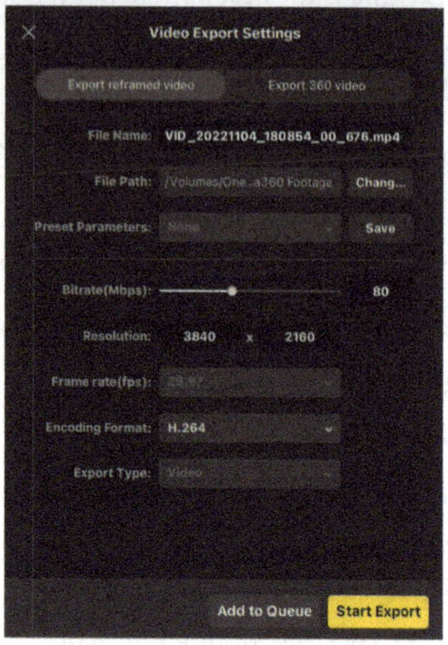

Select the same frame rate you recorded in- for example 29.97 (30) FPS or 59.97 (60) FPS.

The ProRes 422 encoding format outputs the highest quality exports if you are bringing your clip into another editor to mix with other clips, add music, etc.. For clips that you will be sharing without further editing, use H.264 with a Bitrate around 70 for 4k or 25 for 1080p.

You now know how to edit a reframed 360 X3 video clip edited BY YOU in the Insta360 Studio video editing app! Congrats!

TIP: Exporting Multiple Versions of One Edit. If you change the Aspect Ratio after reframing, you can export multiple aspect ratios from a single edit depending on where you plan to upload it. The reframing and keyframes will remain the same. This is really useful if you want to upload the same clip to YouTube and Instagram but in different aspect ratios.

TIP: Creating a video edit. After you have exported your individual reframed clips or 360 clips, you will want to bring the clips into a video editor to mix clips, add music, transitions, and titles. You can also mix your clips with footage from other cameras to create a finished edit. There are lots of options for computer-based video editing apps. DaVinci Resolve is a free, professional level editing app. You can follow the steps given below "How To Output and Edit A 360 Spherical Video" to edit a traditional video. The only difference is when you set up the video settings, you should use a 16:9 aspect ratio, such as 3840 x 2160 for 4k or 1920 x 1080 for 1080p video.

VSDC is another free video editing option with a simpler interface. Other professional editing options include Adobe Rush, Adobe Premiere Pro, or Final Cut Pro.

TIP: Preparing Files For YouTube. When producing videos for YouTube, you can upload either 1080p or 4k (called 2160p on YouTube) resolution videos. The YouTube uploader will automatically detect your video resolution and give viewers resolution options to suit their device.

PRO TIP: Using GoPro FX Reframe in Adobe Premiere Pro® (MAC and Windows). If you are editing in Premiere Pro, you can use a third party plugin called GoPro FX Reframe to reframe your shots. This is optional since you can do all of the reframing using Insta360 Studio. However, some people may prefer to reframe in the Premiere Pro timeline with the main video edit they are working on.

GoPro designed the GoPro FX Reframe Plugin (which is free) for Adobe Premiere Pro® and Adobe After Effects®. This plugin is a powerful tool for reframing your videos, especially since it works with Insta360 X3 footage. The plugin is available for download from the Downloads page on Insta360's website.

If you have Premiere Pro installed on your computer, it will automatically be added as a plug-in.

Since most Premiere Pro users are most likely more experienced, we will give you a quick overview of how to use GoPro FX Reframe with your 360 clips.

1.1 Import. To use GoPro FX Reframe, import your 360 video into Premiere Pro. You can import just one of the two files directly without editing in Insta360 Studio first, but you will get better stitching results by exporting the 360 video from Insat360 Studio first.

1.2 Add GoPro FX Reframe. Then drag the GoPro FX Reframe onto the clip from the Video Effects.

1.3 Use the GoPro FX Reframe controls in the Effect Controls to **match the projection (export size)** to your timeline.

1.4 Add keyframes to the GoPro FX Reframe controls, like you would normally do in a Premiere timeline. You can reframe the shot by manually adjusting the Pan, Tilt, Rotate, Lens Curve and Zoom values in the Effects Control dialog. Or to make visual adjustments using a mouse, click on the GoPro FX Reframe text in the Effects Control Panel. This adds a grid to the media preview screen (called Program Monitor) providing the tools you need to dynamically reframe your video directly in the Program Monitor. The **five basic adjustments are available using these dynamic controls. This diagram shows how to use each control:**

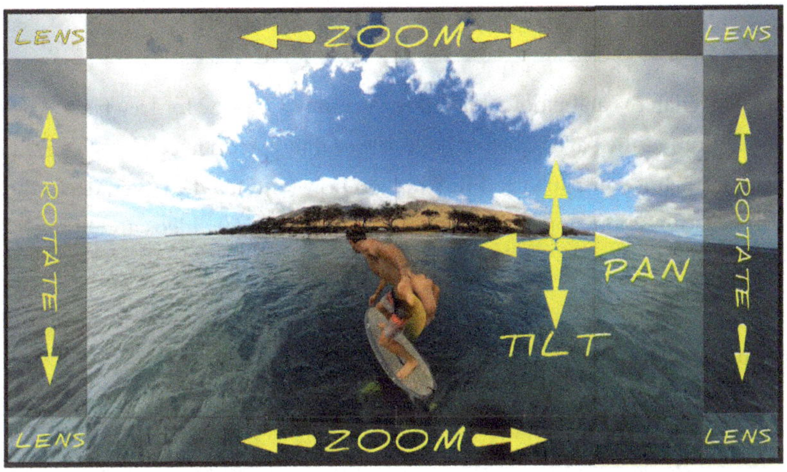

The Dynamic Reframe Controls in the Premiere Program Monitor

1.5 Adjust the transitions between keyframes. After you have finished reframing your video, you will want to adjust the transitions create smoother transitions from keyframe to keyframe. Drag your mouse to highlight all of the keyframe icons in the timeline next to the Effect Controls.

1.6 Color grade and finish editing. After making your adjustments, use Lumetri Color, LUT's (which are preset color correction values), or presets to adjust the color of the clip. Finish editing the clip like you would normally do.

1.7 Preview your edited movie by dragging the Playhead to the beginning of the Timeline. Press the Spacebar to play.

1.8 Select your export settings. When you are happy with your edited video, click File>Export Media. This will bring up the export screen.

Premiere offers lots of export options, including presets for YouTube and Vimeo.

- For this tutorial, let's use the YouTube preset. In Presets, select YouTube1080p HD since we set this tutorial up for a 1080p video. For higher quality export, under the Video settings, change the Bitrate to CBR (longer export time) or VBR 2 Pass (longest export time). Change the Bitrate to around 20Mbps for 1080p. In VR Video, uncheck Video is VR since we are exporting a flat video and don't want to add metadata saying it's a VR video.
- In the future, if you set up your timeline for 4k and want to export 4k video, select YouTube 2160p. Make the same export setting changes as you did for 1080p, but the Bitrate should be set to around 60 Mbps.

1.9 Export the video. In Output name, give your video a name, "Insta360 Reframe Tutorial" for example and specify where you would like to save your video. Click Export. Your video will be ready to share after it renders!!!

NOTE: At the time of publication Final Cut Pro can edit spherical 360 videos, but does not work with the GoPro FX Reframe plugin for reframing.

HOW TO EDIT AND OUTPUT 360 SPHERICAL VIDEOS

Most of the time, you will probably reframe your 360 videos to export as standard 1080p or 4k videos for people to watch. If you want to edit your video as a 360 spherical video, here is how you do it.

Davinci Resolve

We will show you how to edit 360 videos for 360 spherical output using DaVinci Resolve since this is a FREE, powerful editing app which compares to Premiere Pro. (Unfortunately, there is not a reliable 360 reframing plugin for DaVinci Resolve yet.)

DaVinci Resolve is a FREE video editor with most of the features of the industry standard editing apps, such as Adobe Premiere Pro and Final Cut. This lesson was specifically designed to teach you how to export 360 spherical videos and aren't complete DaVinci Resolve tutorials. However, if you want to take it further (even to a professional level) BlackMagic offers free training modules to master this powerful app.

> **TIP:** If you **still want to learn more** about DaVinci Resolve after learning the tips in this step, visit the Training section from the DaVinci Resolve Download page. It's full of helpful, free resources to master DaVinci Resolve, including a free downloadable book in PDF format.

DaVinci Resolve 18 (or higher if available) can be downloaded for Windows or Mac for free from www.blackmagicdesign.com/products/davinciresolve at the bottom of the page. The program is free, but you do need to register. Blackmagic also offers a pro version called Studio for around $300 but you don't need this version now (and maybe never, depending on how far you take your video production). With the free version, you can edit high resolution 4k 360 videos and Single Lens Mode videos without a watermark

If you are using Premiere Pro CC or Final Cut Pro, similar editing techniques can be used and some tips for these programs will be offered throughout the editing section.

Follow these step-by-step instructions to create an Insta360 X3 360 spherical video:

#1- Set Up Your Video Editing Timeline and Add Media

For this lesson, we will create a 4k 360 spherical video since the maximum video size using the free version of DaVinci Resolve is 4k. The X3 is capable is producing 5.7k spherical video.

1.1 Open Davinci Resolve.

1.2 Start A New Project. Double Click New Project and give your project a name. Let's call it "Insta360 360 Tutorial."

1.3 Set Up the Project Settings. First, let's set up the Timeline Settings for our X3 video. The Timeline is where we will assemble the video. In the Davinci Resolve menu at the top bar, select File>Project Settings. There are a lot of setting options that can be changed here, but let's just change the basics to output a 4k 360 video. Under Master Settings>Timeline Format, use these settings:

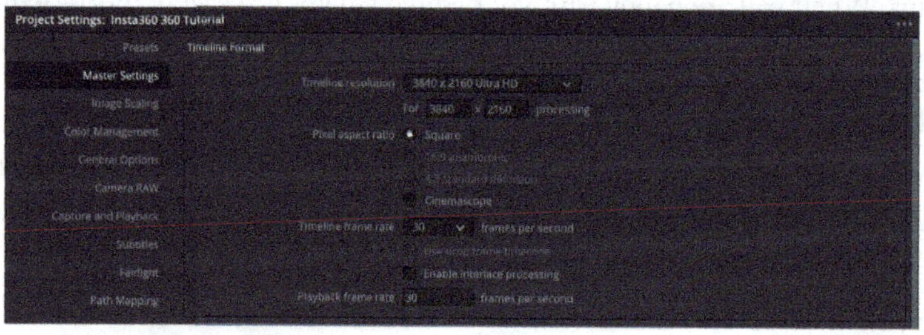

- Timeline resolution: Custom- 3840 x 1920 (360 videos use a 2:1 aspect ratio)
- Timeline frame rate: 30
- Playback frame rate: 30
- If you are using Premiere Pro or the paid version of Davinci, the full 360 file size of 360 video recorded at 5.7k is 5760x2880. 4k 360 video files from the X3 are 3840x1920, which is the same as the 4k resolution in Davinci Resolve.
- For 360 content filmed in 3k, you can upscale the video slightly to 4k without losing much quality but don't zoom in too far.

TIP: Editing a Traditional Video. If you would like to follow these steps to edit a compilation of traditional reframed video shots, use a resolution of 3840 x 2160 for 16:9 4k or 1920 x 1080 for 16:9 1080p.

1.4 Import Media. Drag the two converted video clips from the folder you created into the "Media Pool" area on the left side of the DaVinci screen. If you see the dialog box saying, "The Clips have a different frame rate than the current project settings," select "Don't Change." (We already set up our project's settings for 360 output.)

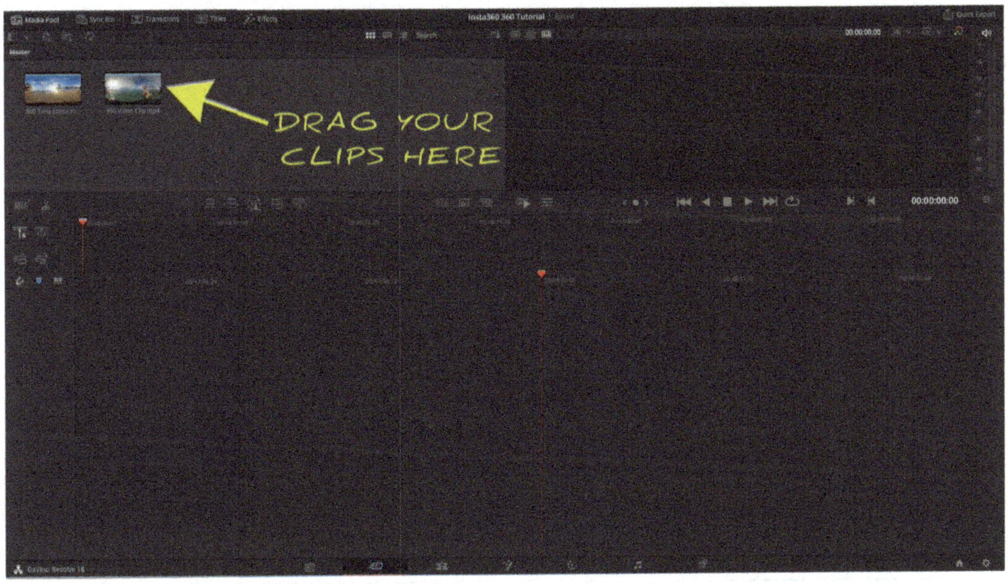

The Media Pool is in the top left corner of DaVinci.

#2- Edit Individual Clips

(All of these edits are done in the Edit tab at the bottom of the screen.)

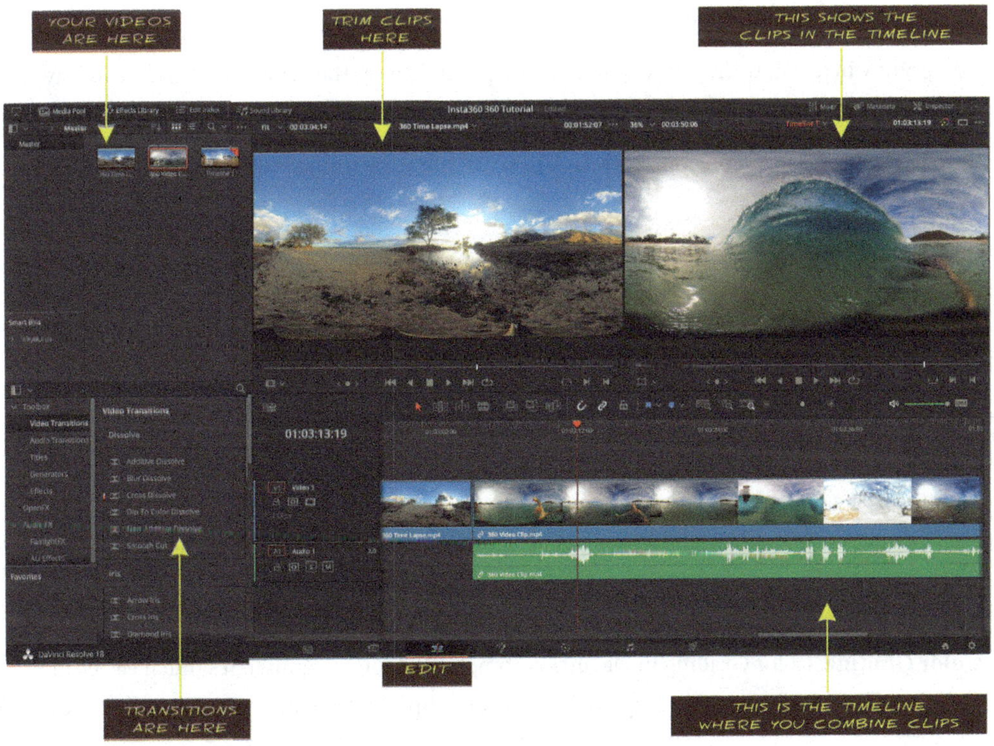

Use this as a quick guide to the DaVinci Editing screen. The Source Viewer is the screen on the left, where you trim your clips.

2.1 Select a Video Clip. In the Media Pool, double click on the first video clip. This will bring a full frame view of the clip into the 'Source Viewer," which is the screen on the left at the top of the Davinci Editing screen. The view will show an equirectangular video. This is what you will use to edit and export the video.

2.2 Adjust the Frame Rate for Slow Motion. If you want a clip to play at normal speed, you don't have to change anything. If you want the clip to play in slow motion, this is where the slow motion frame rates (60 FPS) come into play. By converting the high frame rate to play back at a slower frame rate than recorded (30 FPS), the footage will play back smoothly in slow motion.

To change the frame rate for slow motion for clips recorded at 60 FPS, right click (Ctrl+Click on MAC) over the clip in the Media Pool. Select "Clip Attributes" and change the Video Frame Rate to 30. Clips, or portions of the clip, can still be sped back up in the Timeline to play at regular speed.

You don't need to change the frame rate for clips that will be played back at regular speed.

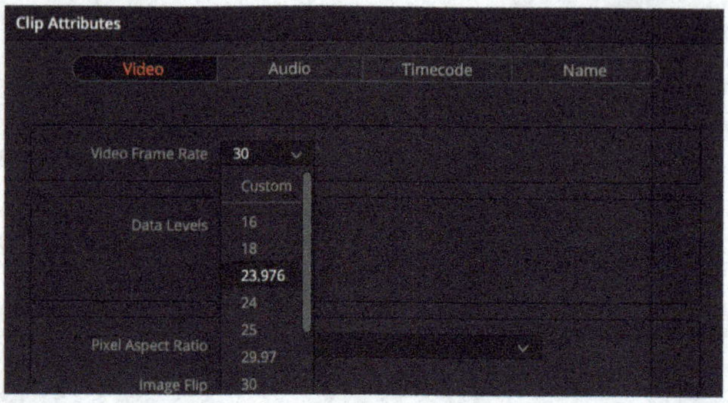

For videos recorded in a Slow Motion frame rate (60FPS), change the video frame rate in the clip attributes to 30 (or 23.976 if you prefer). The clip will play in slow motion when it's added to the timeline in Step 3.4.

TIP: If you drag a clip into the Timeline without adjusting it for slow motion and then reduce the Clip Speed in the Timeline, the clip will play back choppy.

2.3 Trim the clips. Select the In and Out Points, the points where your clip will start and stop, by dragging the Slider located directly below the viewer. Drag the slider to the moment where you want your clip to begin, and press I on your keyboard to set the "In" Point. This will leave off anything before that point.

Drag the slider to the point where you want your clip to end and press O to set the "Out" point. This will leave off anything after that point.

You can also use the left and right arrows on your keyboard to scroll frame by frame. These "In" and "Out" points can also be adjusted when the clip is in the timeline.

2.4 Add the clip to the Timeline. Place your cursor over the video in the Source Viewer and drag the clip to the Timeline below.

TIP: Isolate Video or Audio. You can grab only the video by hovering over the preview in the Source viewer and dragging the filmstrip icon to the timeline.

2.5 Repeat steps 2.1-2.4 for the other clip.

TIP: After the first clip is added to the Timeline, you can drag the next clip either down to the Timeline or to the Timeline Viewer Screen on the right. When you drag it to the right, you will see some of the most common input commands to choose from.

TIP: Color Grading. Color Grading in Davinci is complicated for a beginner. It's much easier in Premiere Pro. To keep it simple, we will cover adjusting the look of your clips (called Color Grading) in the next section.

#3- Add Transitions, Music and Export Your Video

3.1 Add transitions in between the clips for a more professional look. On the top left of DaVinci next to the Media Pool, click on the Effects Library. In the Toolbox, under Video Transitions, select Cross Dissolve and drop it between two clips on the timeline. You can then click on the Transition to adjust the speed of the transition. There are lots of fun transitions to play with, so have fun with this.

3.2 Add a title to the beginning. In the Effects Library toolbox, select Text from Titles and drag it to the empty space (called a track) above your first video clip. You can customize the wording and appearance in the dialogue box on the right. For 360 videos, keep your text relatively small and centered in the frame to avoid distortion when the clip is viewed in 360.

3.3 Add music to create the mood. Drag a song file (usually .mp3) from your computer into the Media Pool. Then drag the song onto the Audio 2 bar of the timeline below the video's audio. You can adjust the length of the song to sync with the video by dragging the end to match the end of your video clips. If you don't want the original audio from the video clips in the background along with the music, you can click the "M" (mute) box in the Timeline to mute the audio on Audio Track 1.

3.4 Preview your edited movie by dragging the Playhead to the beginning of the Timeline. Press the Spacebar to play.

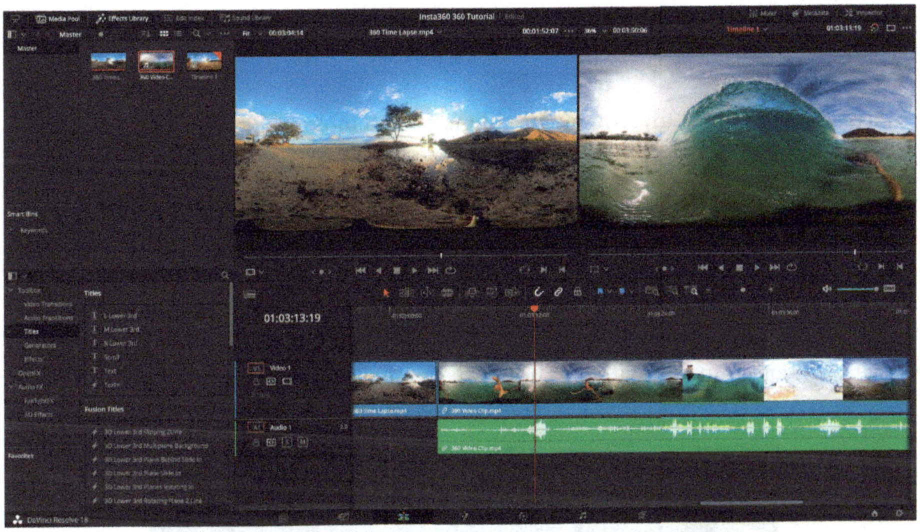

Preview your edited clip in the viewer on the top right.

3.5 Select your export settings. When you are happy with your edited video, click Deliver on the bottom bar of Davinci Resolve (the last icon on the right). This will bring up the export screen.

Davinci Resolve offers lots of export options, including presets for YouTube and Vimeo.

- To export a 360 spherical video, use the Custom tab and under resolution, fill in the resolution (for example 3840 x 1920 for 4k 360 video). Export at 5760x2880 for the full resolution video if you are using Premiere Pro or the paid version of Davinci.

In Filename, give your video a name, "Insta360 360 Tutorial" for example.

TIP: Exporting reframed videos. If you are learning to export regular (flat) videos, export your video at the original resolution you used when you set up the timeline, such as 3840 x 2160 (4k) or 1920 x 1080 (1080).

3.6 Export the video. Click the "Add to Render Queue" button and then the "Render" button on the right side of the screen. This will export your edited equirectangular video.

3.7 Reinsert Metadata (for 360/VR videos only). You may need to reinsert the metadata so your video is recognized as a 360/VR Video. If you edit in Premiere Pro, check "Video is VR" in the export settings to add the metadata. For videos exported from Davinci Resolve, use the Spatial Metadata Injector Tool, which is a free desktop app, to reinsert metadata. You may have to allow permissions, but the app is safe to use. (On a MAC, right click over the app in the Applications icon in the Finder window and select Open from the drop down menu.)

You now have your first edited 360 X3 video ready to view as VR!

TIP: Uploading 360 Videos to YouTube. When uploading 360 spherical videos to YouTube, it's best to upload the highest resolution videos that your camera will produce, which is 5.7k on the Insta360 X3. The YouTube uploader will automatically detect your video resolution and give viewers resolution options to suit their device. The video will take a while to process as a 360 video, so post it as "Unlisted" and go public once you have verified that it's showing in 360. Google recommends adding something like this to the description for your uploaded 360 videos:

"This is a 360 video, so pop on your Google Cardboard or VR Headset to be totally immersed in my world! No headset? No problem. Move your mobile phone around and catch the total 360 experience."

<u>GO DEEPER</u>

MORE EDITING TECHNIQUES. Once you have some editing practice and are comfortable with the basics of video editing, check out the options below for some extra effects.

COLOR GRADING

Color Grading is the video editing term for **adjusting your video's overall look.** Color grading can get pretty in depth, but let's just start with a few of the basic adjustments you can make to improve your video's appearance. DaVinci Resolve is a powerful color grading tool used for many major movies. The Color Grade tab at the bottom of the screen brings up a very overwhelming looking screen. However, we will just touch on a few of the basics and you should be fine. Blackmagic offers a 2-hour training video just on adjusting colors for those of you who are interested in learning more at a later time.

Three of the most useful color grading tools are the Lift (dark tones and shadows), Gain (highlights), and Gamma (midtones) wheels, but these are way too advanced to start with. For now, you can make some more basic adjustments to your video's appearance by **following these steps.**

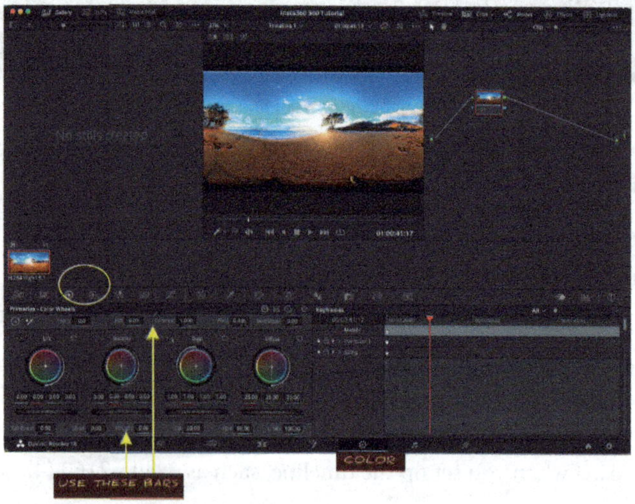

1. Select the "Color" icon on the bottom of DaVinci Resolve to go to the Color page.

2. Click on the thumbnail of the clip you want to edit.

3. In the middle left of the screen, make sure the Color Wheels are shown. If not, select the circle icon below the thumbnails on the left.

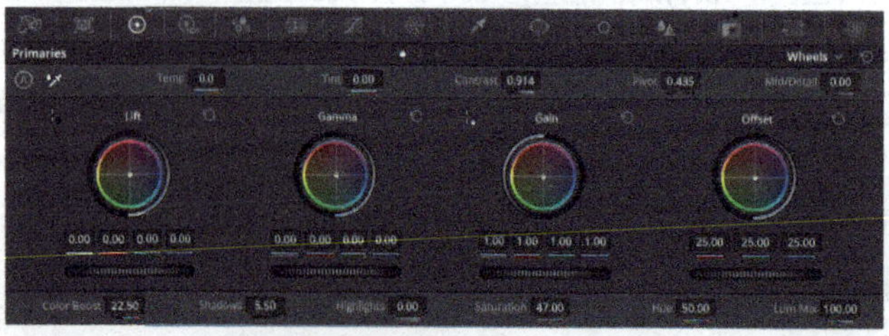

Just use the adjustments on the top and bottom bars of this Color Wheel box.

4. On the bottom bar, you will see a dropper icon, the letter "A" in a box and the numbers 1 and 2. Press the "A" to view an automatic color adjustment. You can use this as a starting point, or you can select Edit>Undo to return to the original settings.

5. To the right of the "A", the 1 and 2 buttons access some of the more traditional adjustment tools such as Temperature, Color Boost, Shadows, Highlights, Contrast and Saturation. If you hover the mouse over them, you can slide either way to adjust the values and preview how it affects your video image. Be careful not to get too carried away because overediting can make your image look doctored, especially when using the Saturation slider. Contrast, Color Boost, Highlights and Shadows are the most useful to start with. These changes shouldn't affect your stitch lines, but make sure to check when you preview your video in 360.

> **TIP: Preset Color Adjustment Tools.** As you become more experienced, you will develop a style and settings that appeal to you. You may also want to download some LUT's, which are color grading presets you can use to achieve a uniform look throughout your videos, but that is something to remember for the future. Powergrades are another DaVinci Resolve color editing preset tool to look into as you become more experienced.

ULTRA SLOW MOTION

Changing the playback speed of your videos is **one of the best ways to create exciting, captivating videos.** Since the X3 doesn't offer super high frame rate settings for ultra-slow motion, we will have to use some **tricks to digitally create slow motion.**

You can edit ultra-slow motion segments to reframed traditional video or to spherical 360 videos.

To create Ultra Slow Motion, DaVinci Resolve has a great tool called **Optical Flow** you can use to play back your videos far slower than the actual frame rates would allow.

As you learned previously, to achieve true slow motion, you are limited to the frames that were actually recorded in the original video. For example, when you play back a video recorded at 60 FPS at 30 FPS, you can only slow down the video 2x to maintain smooth video footage. That is only 50% of the original speed.

One of the great benefits of using DaVinci Resolve to edit X3 videos is you can utilize an effect called Optical Flow to edit portions of your clips into Ultra Slow Motion. Optical Flow can also be applied in Premiere Pro or Final Cut using a similar technique, but DaVinci generally produces better results.

With Optical Flow, you can slow your footage down to 3%-10% to achieve super ultra-slow motion far slower than the original frame rate would allow. This creates an almost paused effect which is perfect for accentuating those really beautiful, extreme or funny moments.

This is also a huge benefit when editing X3 videos because of the lack of high resolution super slow motion settings.

However, not all is perfect with Optical Flow. To achieve super slow motion, the program makes up frames that don't exist by analyzing the frame before and the frame after and then fills in the space with digitally-created frames. For most scenes, this works really well, but for some scenes, the results are less than perfect.

Consider the following tips when you want to edit your clips into ultra-slow motion:

- When possible in Single Lens Mode, record at 60 frames per second so the program doesn't have to create so many frames. The fewer frames the program has to digitally create, the better your chances for an effective result. However, Optical Flow can be used with footage recorded at any frame rate.

- **Choose the right scenes** to apply Optical Flow to. Clean backgrounds and simple scenes work better because they are more predictable. Busy scenes with lots of background elements, such as bushes or water, make it harder for the program to imagine what the missing frames should have looked like. You can try Optical Flow with any clips, but if the results come out strange, the background could be the problem.

- Optical Flow can be applied to any video that has been slowed down beyond 24 frames per second to prevent choppy video.

You've come a long way since we first met because this is getting into some more complicated editing. But we will walk you through step by step to add Ultra Slow Motion to your clips:

1. In the Edit Tab, drag the clip (which you already exported from Insta360 Studio) into the Timeline. Make sure the Clip Attributes are adjusted to play back in slow motion to start with the slowest motion possible.

2. Right click (Ctrl+Click Mac) over the clip in the Timeline and select "Retime Controls." You will see 100% and a small arrow show up on the bar below the clip in the Timeline.

3. Add Speed Points to a Short Section of the Clip. Drag the Playhead to the beginning point where you want to change speeds. Click the small arrow next to 100% and select "Add Speed Point."

4. Drag the Playhead to the next point where you want to change speeds. Click the small arrow next to 100% and select "Add Speed Point."

5. Adjust the speed of the clip by clicking the arrow in between the points you just marked. For Ultra Slow Motion using Optical Flow, reduce the Speed to 10%.

6. To fine tune the speed and smooth out the transitions between speeds, use the Retime Curve. Place your cursor over the clip, Right Click (Ctrl+Click Mac) and select Retime Curve. A new bar will show up below the clip. Click the small arrow next to the words "Retime Frame" and select "Retime Speed". You can fine tune the speed by moving the level up and down.

7. Important Step! for Optical Flow. To add the Optical Flow, Open the Inspector Tab at the top right. Double click Retime and Scaling and then select:
Retime Process: Optical Flow
Motion Estimation: Standard Better (This setting depends on the speed of your computer and desired quality. As you move down the menu, the quality increases, but so does the render time.) The last option, Speed Warp, is only available on the Paid Version of Davinci Studio, but the other settings produce great results.

> **NOTE:** Depending on your computer, exporting with Optical Flow applied takes longer because Davinci Resolve has to analyze and create new frames, so make sure you only apply Optical Flow to the clips that need the extra frames.

• If you want to reframe your 360 video after adding super slow motion, export the video as a 360 file. Then bring it back into Insta360 Studio. Studio will recognize it as a 360 video because of the proportions.

OTHER QUICK FIXES

Sometimes footage doesn't come out like we hope for. Fortunately, some things can be corrected after the fact. Here are a few common filming mistakes you can fix when editing

• **Underexposed shots.** If you filmed in low light, or with a max ISO that was too low for the lighting conditions, your footage may be darker than desired. You can brighten the exposure in any editing app to lighten the shots, but make sure you don't go too far because it may bring out noise in your shots. In Davinci Resolve, the easiest way to adjust the exposure is to adjust the Shadows and Highlights in the bottom bar that you were using for color grading. It can get much more involved, but this is a good place to make simple adjustments. In Premiere, use the Exposure slider in the Lumetri color panel.

• **Audio.** The easiest option of course is to use the X3's in-camera audio options to record the best possible audio on the spot. If you already recorded the footage and the audio is less than par, you can first try to correct it using audio adjustments in DaVinci Resolve or another editing app. The last option may be to mute the track's original audio and record a voiceover, use a separate sound effect audio track, or use that clip with music as the audio.

EDITING A 360 TIME LAPSE VIDEO

Time lapses are fun, creative additions to any video, and they can give your viewers a great understanding of the overall scene. As you learned in Step 2, the X3 records 8k time lapses using 360 Time Lapse Mode. This is the best option for recording scenic time lapses because of the high resolution and they are automatically formatted into a 360 video format. You can then export the time lapse as a 360 experience or a high quality reframed video.

TimeShift Mode can be used for hyperlapses or action time lapses, but you will usually find better results with recording regular speed 360 video and speeding up the clips using the Insta360 App or Insta360 Studio. The TimeShift tool in the apps gives you more creative control over speed and motion blur.

Just remember that frames are much further spaced out in a time lapse. A video recorded at 30 FPS needs to be sped up 1500% to equal a .5 second interval. This will eliminate the frames in between to create a time lapse effect. The fastest speed available in Insta360's apps is 64x the original speed.

Editing Tips:

- Work with the speed. **Try adjusting the speed** to get the look or length you are going for. Speed changes add a lot of drama to time lapse videos. Speed adjustments can be made directly in Insta360 Studio or the Insta360 App.

- **Use keyframes to add extra movement,** such as panning or zooming, to your time lapse. 360 content really gives you a lot of flexibility to move around the scene. Long drawn out movements sync well with the fast action of a time lapse giving the reader a calm sense of bewilderment.

- Depending on your computer's speed, your time lapse may not play back smoothly until you export it. The preview doesn't show all of the frames, so it will look choppier than the final exported video. **Export your video from Insta360 Studio or the App to view the full effect of your time lapse.**

PULLING FRAME GRABS

A frame grab is **a still image (photo) taken from a video clip.** You learned how to pull still frames from video using the Snapshot icon in the apps. With the high video image quality on the Insta360 X3, a still image taken from 5.7k video produces high quality photos to work with. And with at least 24 frames per second, you can really utilize your video clips to extract amazing still images. Frame grabs pulled from video have the same resolution as the video file, so a higher resolution video produces a larger still image. These still frames can then be edited just like a regular photo, like you will learn about next.

Extracting a **flat frame grab** from either 360 or standard video is easy to do directly in Insta360's apps. Simply reframe your shot and use the Snapshot tool to pull a still photo of the view you are seeing.

To **save a 360 image** from a 360 video in the Insta360 App, simply use the Snapshot Tool and select "Save a 360 image." The 360 photo will save to your device's Photo library.

To pull a **360 Snapshot** from a 360 video in **Insta360 Studio**, you need to follow a couple of steps. With your 360 video open in Insta360 Studio, make sure you are in "360 View" at the top of the screen. Then, under the FOV options on the video preview, select "Equirectangular Panorama." Now, scroll to the frame you want to capture and click the Take Snapshot Icon. Your Snapshot will now be ready to view in 360.

> **TIP:** Adjust colors before saving the Snapshot on the Insta360 App. Since the Insta360 App offers color grading tools, it's easiest to adjust the colors before exporting your Snapshot.

PHOTO EDITING

When it comes to 360 photography, there are lots of things you can do with the Insta360 X3! There are tons of ways to edit photos and you will eventually find your own style. This section will teach you how to make corrections to your photos as well as use editing techniques to **give your Insta360 X3 photos or frame grabs the extra pizazz they need** after coming straight from the camera. This section is written to help you process photos to make them look their best!

As with video, you can output your photos to be viewed in either a 360 spherical format or as a "reframed" flat photo. We will first teach you specific steps for editing 360 photos for a spherical output since they require their own workflow. We will then show you some of the best and easiest editing techniques for flat "traditional" photos, such as Single Lens Mode photos or reframed photos.

360 PHOTO EDITING WORKFLOW

For virtual tours and scenic 360 photos, you will need to edit and output 360 photos for spherical viewing. Because they are being viewed seamlessly as a spherical image, 360 photos need their own special techniques to maintain a seamless image. Even though 360 photos are often presentable out of the camera, they do need common editing adjustments such as color correction, erasing unwanted objects, and fine tuning the quality. Follow the steps below for an easy workflow and to learn about the best apps for editing your 360 photos.

EXPORTING 360 PHOTOS

360 photo editing is available for any photo taken using 360 Photo Mode or any frame grab pulled from a video recorded in a 360 Video Modes. We taught you how to pull a frame grab from 360 video in the previous section.

360 photos are saved in an .insp format and need to be opened in the Insta360 App or Studio before opening in another editing app.

METADATA

After you edit using the recommended apps, the photos should be saved with the photo's metadata telling the host that the photo is a 360 Spherical Photo. **If the metadata is missing, the photo will not display properly in 360.** If you pulled a still 360 spherical photo from a video, that file may also lack the metadata needed for 360 editing.

Usually, the metadata will be stored with the photo, even if you pulled a still photo from video. If the metadata is missing, use the following tools. **On a phone or tablet,** you can reinject the metadata using Meta360, which is a paid mobile app for reinjecting metadata. All you have to do it import the photo, inject the metadata and export the photo. It's quick and simple. **On a computer,** use Exif Fixer (by panoramaphotographer), which is a free desktop app.

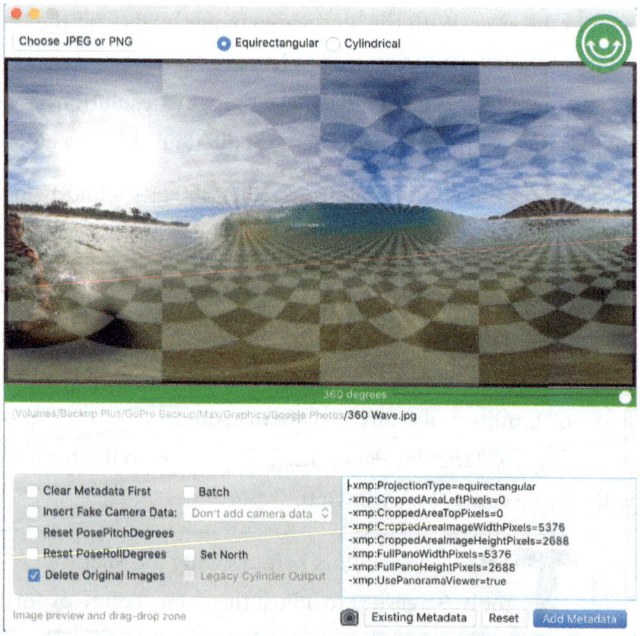

The free Exif Fixer App for reinjecting photo metadata on a desktop.

BASIC EDITING AND COLOR CORRECTION

It's important to use the right editing apps for your photos that will maintain a seamless photo when viewed spherically. When you edit a 360 Spherical photo, it will be viewed as an equirectangular photo. This can be odd to work with at first, but as you edit, remember that when viewed in 360, the edges will be viewed seamlessly.

Unfortunately, Insta360's Apps don't have color grading tools for photos, and the third party options are limited. Google Photos is free and offers a few limited color correcting tools. The best full-feature photo editor to use for editing 360 photos is called Affinity Photo. Affinity Photo is reasonably priced and only requires a single payment, compared to Photoshop which requires a subscription. (Photoshop used the be the best choice, but they recently removed the 360 photo-editing tools.)

TIP: 360 Raw (.dng) Photos. Raw photos must be edited using Affinity Photo after exporting the raw 360 photo file from Insta360 Studio.

Use the following apps when editing your 360 photos for seamless 360-aware editing:

Google Photos (free)

Google Photos is **a free desktop editing tool for editing your 360 photos.** Google Photos can also be accessed on mobile devices, but the mobile app doesn't offer the same editing tools as the desktop app. Edits on a mobile device will have to be performed in Snapseed (which is Google's mobile photo editing app) or another mobile editing app such as Lightroom Mobile.

Google Photos automatically recognizes an Insta360 X3 spherical photo (as long as it has the metadata) so you can view it as a 360 image. When you make edits, the app is aware of the edges so your stitch lines will remain unseen. The editing tools are limited in Google Photos, but it's a great free resource to use for basic color correction.

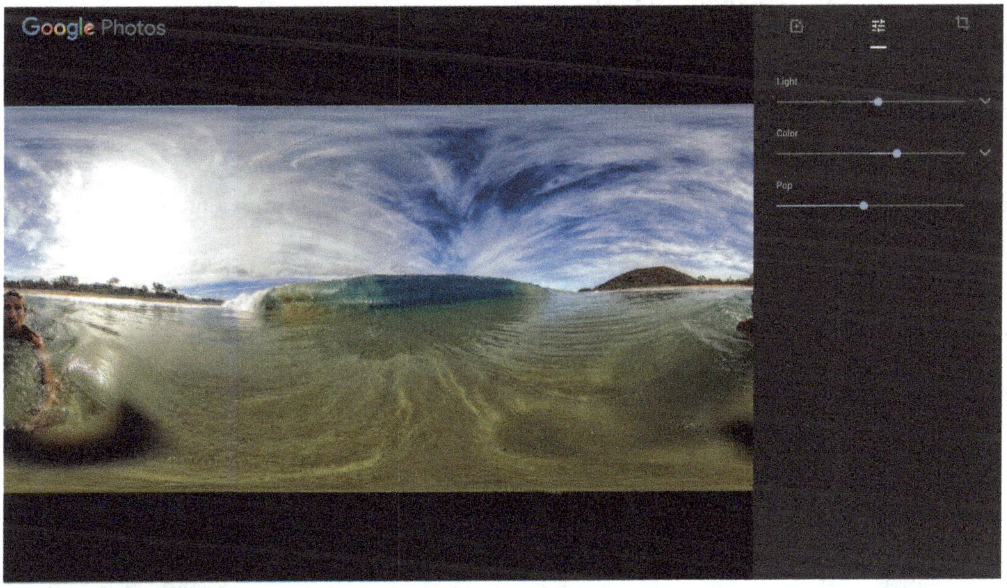

To Use Google Photos:

1. Open your 360 spherical photo in Google Photos. The photo will show up as a 360 spherical photo.

2. Tap on the Slider edit icon and the photo will open in the editing window as an equirectangular photo.

3. Add a color filter or use the adjustment tool to manually adjust Light, Color and Pop. Avoid using the crop tool because this will ruin the seamless edges of your photo.

- Unfortunately, Google Photos does not offer a Cloning tool yet, so you will have to turn to another editing app such as Photoshop or a mobile app for cloning.

- If you want to edit your 360 photos on a mobile device using Lightroom Mobile or Snapseed, we will cover the steps for that too.

Affinity Photo (not free, but reasonable)

Since Photoshop removed their tools for editing spherical photos, Affinity Photo now has **the best tools for complete editing of your spherical 360 photos.** You can edit your images from start to finish using Affinity on a computer for Windows/Mac or an iPad (iOS). If you are editing real estate photos for a virtual tour or scenic photos for display prints, it will be worth your investment to buy Affinity Photo.

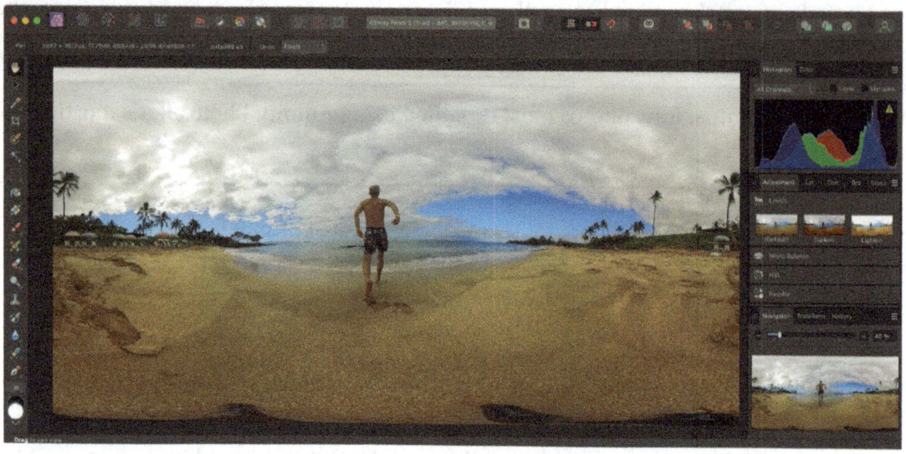

The Original Photo when viewed in Affinity at the full 2:1 ratio.

The following steps provide you with a basic workflow for editing 360 photos in Affinity Photo:

1. First **open and export the photo** from Insta360 Studio (computer) or Insta360 App (iPad) as a .jpeg or .dng (for RAW photos).

2. In Affinity Photo, **open the spherical .JPEG or .dng image** (File>Open). We will work on the computer version.

3. 360 Equirectangular photos can be worked on in the original 2:1 ratio, or as you would view the projected image in 360.

4. First, **make universal adjustments** while viewing the image as an equirectangular image using the Adjustment Tools. To open the Adjustment tools, select Window>Adjustment.

5. The **most useful adjustments** in the Adjustment panel include Temperature, Exposure, Contrast, Clarity and Vibrance., but you can fine tune the overall look of your photo using these tools.

6. After you are satisfied with the overall look, in the Layers Panel on the right, **click on the Pixel Layer** (the layer with your image on it.) Or **if you are done with the universal adjustments, merge the layers.** From the top menu, select Layer>Merge Visible.

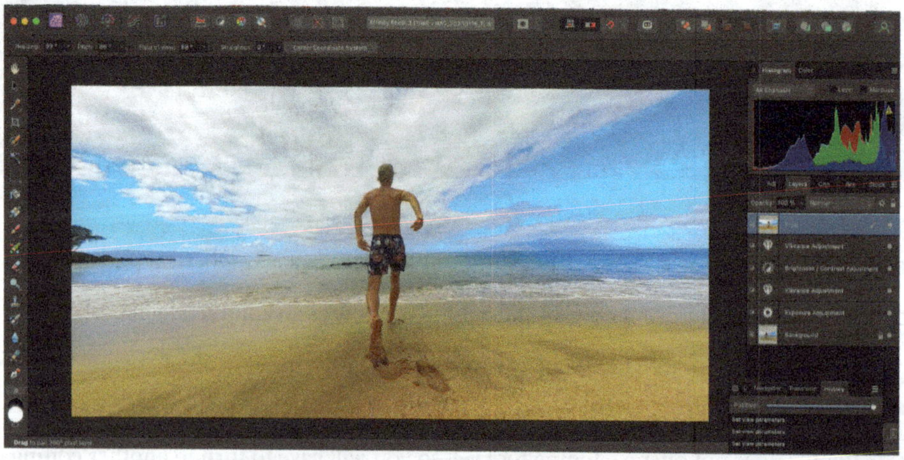

The same photo in Equirectangular Projection Mode.

7. Next, **make local edits to specific areas of the photo** with your image as a projected image (how you would view it in 360). Editing out unwanted objects is much easier when viewing the image as a spherical image. With your image layer selected, from

the Layer Menu, **choose Live Projection>Equirectangular Projection.**

8. You can now pan around your image and **view it in 360.** At the top of the window, adjust the FOV to zoom in or out on the image.

9. Navigate around the panorama **using the tools needed to make local adjustments** (adjustments to specific areas), such as cloning out unwanted objects such as the tripod feet (using the Clone Tool), and dodging (lighten) and burning (darken) to lighten or darken specific areas of the photo. When viewed in 3D, objects such as monopod feet are easy to clone out of the image.

IMPORTANT NOTE: When you switch tools when panning in Live Projection, **you will temporarily leave the Edit Live Projection Tool.** To pan around the image again, go back to the Layer Menu and choose Live Projection>Edit Live Projection. Or select the Move Tool and then click Edit Live Projection at the top of the editing window.

10. To **return to an Equirectangular View** when finished, select your image Layer in the Layers Panel, then from the Layer menu, choose Live Projection>Remove Projection.

11. Lastly, **flatten your image** (Layer>Merge Visible), and **export the 360 photo** (File>Export). Select your desired file type, make sure "Embed Metadata" is selected and save your finished photo.

CLONING

Removing unwanted objects is much easier in a photo than in a video. In 360 spherical photos, **cloning is especially useful to remove the tripod/monopod feet** if your camera was mounted on a monopod. If you are trying to clone out the monopod feet when the photo is begin displayed as an equirectangular image, the tripod feet will appear spread out along the bottom of the screen making this a hard area to clone. This point directly below the camera is called the nadir. The nadir is typically the area that needs cloning.

There are several techniques to use for cloning objects out of your 360 photos.

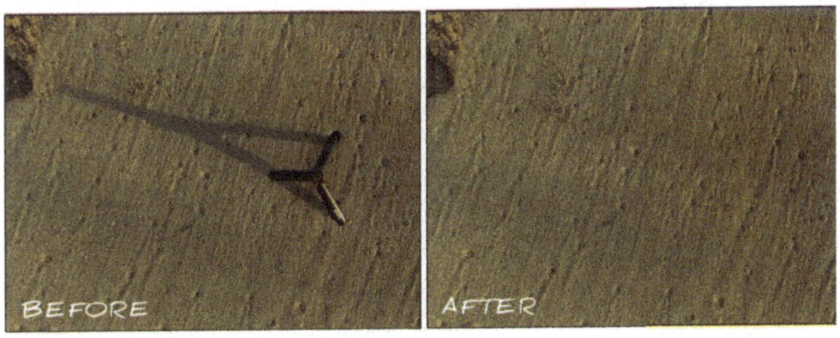

The Clone tool in the Equirectangular Projection View of Affinity Photo is one of the best ways to remove unwanted elements in a photo because you can edit in 360, rather than equirectangular. The monopod base and shadow were removed in the image on the right as an example of what can be done quickly and easily.

In Affinity Photo on a computer, use the clone tool when you are in Edit Live Projection View as you learned in the previous step. This brings the tripod feet together how you would see them in real life, making them easy to clone out.

The online 360 editor Kuula also has an easy-to-use tool for removing the tripod from your shots.

Most of the mobile apps, including Snapseed or Photoshop Express let you clone out unwanted areas using the Healing Brush on your touch screen.

However, it can be tricky to fine-tune cloning on your device especially with an equirectangular 360 photo. The tripod feet end up being spread out along the base of the equirectangular photo.

The best paid app for removing monopod feet is Touch Retouch (iOS/Android). This app allows you to edit in a 360 view (similar to Affinity Photo) to bring the tripod feet back together in an equirectangular view, allowing you to clone out the tripod all together.

ANIMATING YOUR 360 PHOTOS

Animating the reframing adjustments on a 360 photo really brings the image to life. If you want to **animate a 360 photo** (like you did using keyframes for a 360 video on the Insta360 App), open the photo in the Insta360 App. At the top of the screen, select Animate and choose the preanimated options. Editing tools are also available so you can fine tune your image before exporting your animated photo. You can then share this photo with animation on Instagram or host it on your website/blog. (In Step 6, we will show you how to host 360 photos to a site.)

For more variety, the Theta+ App also offers some great 360 photo animation options.

MORE 360 PHOTO APPS

Circular Tiny Planet Editor. This is a paid app, but the tiny planet effects are definitely worth it if you want to have some fun creating them. You can add multiple layers of special effects, such as starry skies, clouds or birds. It's a great app for getting creative with your tiny planet photos.

EDITING REFRAMED PHOTOS AND SINGLE LENS MODE PHOTOS

Reframed and Single Lens Mode photos are flat photos that don't require the same specialized workflow as 360 spherical photos. These are edited like any other photos from your phone or another camera. However, editing these photos to improve their look is equally important. We will use Lightroom Mobile for editing these photos on your phone or tablet since high quality photo editing can be done using your mobile device. Similar techniques can be used if you are using Photoshop or Gimp (a free desktop photo editing app) on a computer.

Editing flat, "reframed" photos requires a similar process to editing your 360 photos except for a few of the steps. Now let's cover a few tips specifically for editing your traditional photos.

BASIC EDITING (Traditional/Reframed Photos)

Let's start with the basic editing that should be done for any photo. These little adjustments make a big impact. **We are going to use Lightroom Mobile for this tutorial** since it's free and it offers a bit more control over your images than Snapseed. Although we are using Lightroom Mobile for this tutorial, similar adjustments can be made on most photo-editing programs, both desktop and mobile.

Import your photos

After you have transferred photos onto your phone, open Lightroom Mobile and select a photo to work with.

Applying Presets/Filters

Under the Presets option, scroll through the presets to see how they affect the appearance of your photos. This is like adding filters in other mobile editing apps. With experience, you will know your favorite go-to's. Or you may decide that you prefer to manually adjust the appearance of your photos.

Priime is another free mobile app with some amazing filters for your photos.

Color Correction

Straight out of your camera, your photos may look a bit dull. It's easy to correct the appearance of your photos in a few simple steps to add more color and life. These adjustments are made using the Light and Color tools in Lightroom Mobile, but every decent photo-editing app has similar features. In Snapseed, they are found in Tune for JPEGs.

1. Use the Light tool to adjust Exposure, Contrast, Whites and Blacks. Exposure and Contrast are typically the most impactful adjustments to fine tune the tone of your photo.

2. Using the Color tool, the Vibrance and Saturation adjustments will add more impact to your images. When adjusting Vibrance and Saturation, add color until it looks oversaturated and then back it down a bit so the image doesn't look over edited. You will rarely go beyond +25 for Saturation, depending on the photo. If you took the photo using a manual White Balance setting, you may need to adjust White Balance or Temperature. If you used Auto White Balance, these most likely won't need to be adjusted.

TIP: Local Adjustments. If you choose to use local adjustments (for a specific area of the frame) for a 360 photo, avoid editing the areas around the edge because these adjustments could make the stitch lines visible.

3. Lastly, after other adjustments are made, use the Clarity slider in the Effects tool if the image needs more contrast and punch. The Clarity slider boosts the midtones of a photo.

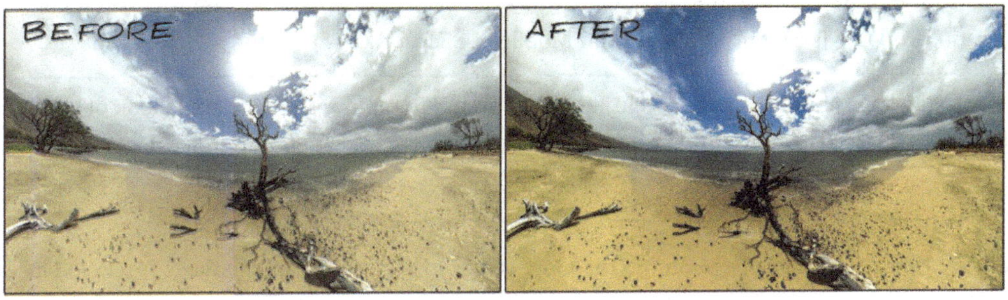

The Original photo is on the left. The image on the right was adjusted using the basic editing adjustments shown above. Here are the adjustments made: Exposure -.16, Contrast +9, Whites +38, Highlights +12, Shadows -12, Whites +21, Blacks -22, Vibrance +29, Saturation +9, Clarity +29.

These same adjustments can also be made using the Develop Module in Lightroom.

REMOVING FISHEYE

There is an easy way to correct lens distortion in your photos if they need a more linear perspective.

Using a lens correction adjustment will cause some loss of the content around the edges of your photos, so make sure you don't lose anything you feel is important to the composition. Also, keep an eye on objects around the edge of the photo to make sure they don't become too stretched out in appearance.

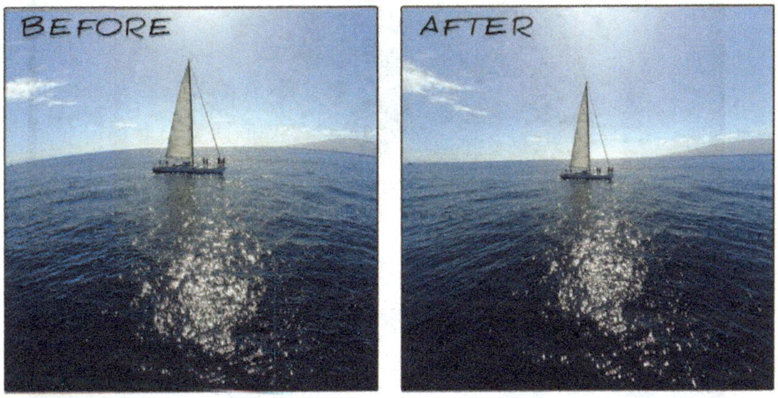

Lens corrections were made to the image on the right to remove the fisheye effect.

The best way to remove the fisheye effect on a mobile app is to use SKRWT, which is a paid app (about $1.99US), but it's one of the best available:

1. Import your photo into SKRWT and select the curved edges icon on the bottom right.

2. Select the GoPro tab on the right. Slide the bottom bar to the positive numbers until you are happy with the results. Between 10-15 is typically a good range to remove fisheye from a photo taken in Wide FOV.

3. Save your image to your phone and make any additional edits using Lightroom Mobile or Snapseed.

If you make the Lens Correction adjustment using Lightroom on your computer, you can also fine tune the amount of distortion removed using the Distortion slider. You can also make this correction in Photoshop using Filter>Lens Correction. Under the Custom tab, adjust the Geometric Distortion slider to remove the fisheye effect.

CROPPING

Cropping can be used to either change the aspect ratio of a photo or to adjust the composition. Since the Insta360 App offers several aspect ratios to choose from when you export your photos, you probably already cropped your image. You also may want to crop your reframed photos into a panoramic aspect ratio, which looks impressive at a 3:1 ratio. If you want to crop further or change the aspect ratio, cropping is easy using Lightroom Mobile or Snapseed.

> **TIP:** Cropping your photo is another technique you can use to remove the wide angle look of your photo. By cropping the edges of the frame and keeping the center of the image, you can remove the areas with the most distortion.

Cropping reduces the number of pixels in a photo, so the more you crop, the more original image pixels you lose. Reframed photos from 360 spherical video are about 16 MP. The high resolution of these photos gives you lots pixels to work with when cropping. If you are using the photo for Instagram or other mobile media, you don't need much resolution so feel free to be bold.

FINISH IT OFF

Once you have finished editing your traditional photos, you can finish them off by adding a few final touches.

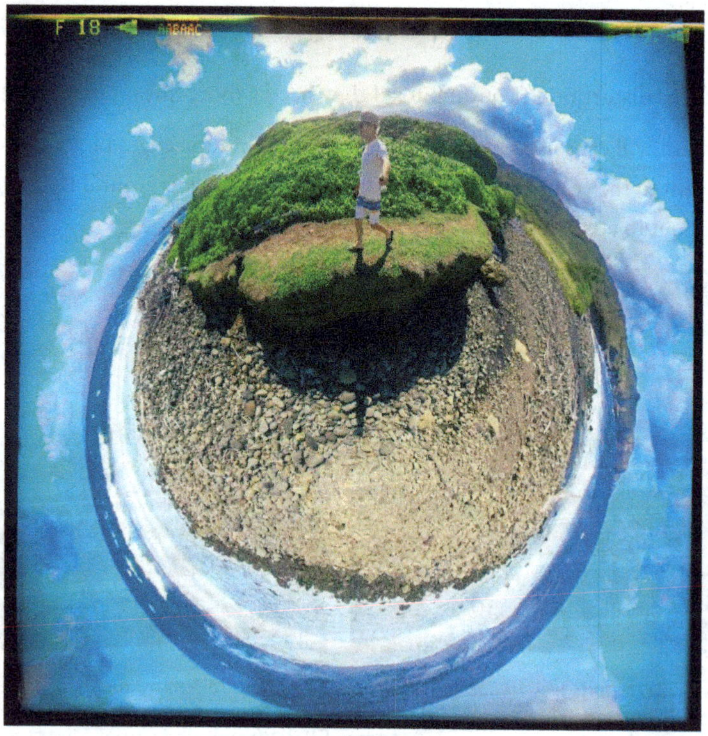

The photo was cropped to a 1:1 (square) aspect ratio and a Vintage Instant filter was added using Lightroom Mobile. A frame was added in Snapseed. These types of filters and effects can easily be added to your photos to make them more fun and stylish!

A vignette adds a dark area around the edges of the frame, or you can choose a stylish edge, such as a film emulsion. Lightroom Mobile has vignette controls. Snapseed and Photoshop Express both have some good edge and frame options.

You can also add some fun lighting effects. Lens Distortions and Pixlr are free apps with lens flares and other effects that are easy to preview and add to your photos.

SPLIT A PANORAMA FOR INSTAGRAM

Want to add a swipeable, multi-panel photo to your Instagram? If you used InstaPano or cropped one of your images to a panoramic aspect ratio, there are a few available apps to easily split the panorama into three (or more) equal squares. When you upload this file to Instagram, viewers can swipe to view the panorama in a series of seamless photos. For more attention on the Instagram search page, post the full panoramic photo first, followed by the sliced pano.

The best apps for creating a swipeable pano are Swipeable (iOS), Unsquared (iOS/Android), Panoragram (iOS), and InSwipe (Android).

Now that you've edited your photos and videos like a pro, you are ready to move on to Step 6 and show them to people!

STEP SIX

SHARE IT

Get Your Vision Out There For People To See

Now that you've filmed and created your media, it's time to share the love. With so many options for your final pieces, you may wonder how to view and share your edited photos and videos.

As you film more and more, you will likely run into the scenario where you filmed somebody doing something awesome, beautiful, creative, or entrepreneurial that you want to share.

Read on to discover the many ways to get your media out there for friends, family and maybe the world to see.

SHARING FILES

SHARING THROUGH THE INSTA360 APP

The Insta360 App enables you to share photos and videos directly from the App. You can share files that are currently on your camera's microSD card or files you have downloaded to your device.

With the amazing editing tools on the Insta360 App, it's super easy to edit and share quick video edits and photos using the Insta360 App. You can share your edits directly from the Insta360 App to social media (such as Facebook, Instagram, etc.) or share with your friends via text messages. **Open the media, tap the Export icon. After exporting, select how you want to share and it's easy as that.**

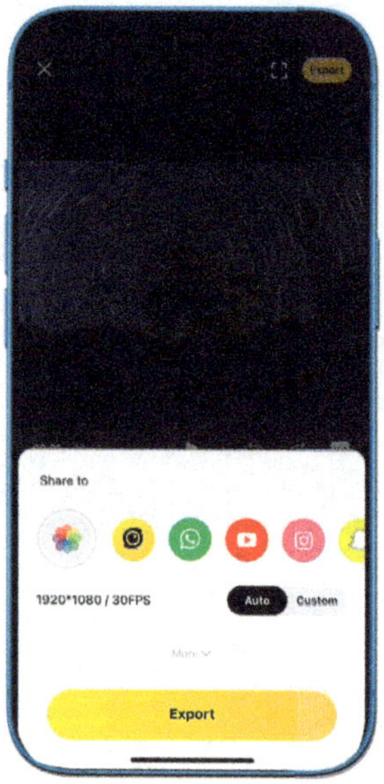

If you want to **share 360 files** with other people who are familiar with 360 editing like you are now, send them the original 360 files to create their own edits. It's amazing how different people will edit footage in such unique ways. For smaller files, you can share the 360 files directly from the Insta360 App. For larger file sharing on mobile devices, after downloading the 360 files to your phone, use WeTransfer to send a download link to the file. Or if you are both on iOS and near to each other AirDrop is the easiest. Use Nearby Sharing for sharing on two Android devices.

If the files are already on your computer and you still want to share them to a phone or tablet, on a Mac, right click (Ctrl+Click) and select Share to select how you want to send the file (Messages, email, etc.) On a PC, select the file in its folder and use the Share Button on the Share tab to share the file. This works for smaller files. Next, we will give you options for large files.

SENDING LARGE FILES

If you were out filming and captured some shareworthy video footage of your friend or even a stranger who was out there shredding, you might be struggling to figure out how to share the large video files. Even a short video file is too big for most email accounts, so you need to find a better way to share files. Dragging the file over to a USB flash drive and loading it onto your friend's computer is one rather old-school, but effective, way of sharing files.

If you don't have the convenience of physically sharing the file, there are a few free ways to send large files. (These are the current size limits at the time of publication.)

WeTransfer is the best free option because it lets you share files up to 2GB as often as you like without signing up for any plans. All you need to do is to enter your email and the recipient's email address and link the file. The transfer can be completed from a desktop or mobile device.

Hightail (formerly YouSendIt) offers a free plan that lets you send files up to 100MB, which isn't much if you are sharing 5.7k 360 files, but it could be sufficient for short clips depending on the resolution. You can send much larger files with the paid plans.

POSTING VIDEOS AND PHOTOS

If you want to provide a way for people to view your edited photos and videos online, there are lots of options available.

360 VIDEO SHARING

If you want to share your videos as a 360 spherical video, you have a few great free sharing options for hosting your videos with the ability to view them in 360. All of these, except for Veer can also be used for hosting your wonderfully edited HD videos as well.

YouTube - *Most popular. Host 360 and Single Lens Videos in Long and Short Form.*
Create a channel and upload your videos for private or public viewing. The Insta360 App provides access for uploading your edited videos directly from the Insta360 App. Or you can upload your videos from a computer or device library.

YouTube allows you to upload 360 video content or standard HD video content for free with a wide range of sharing options. You already learned how to format your 360 videos for YouTube upload in Step 5- Creation.

For long form YouTube videos, a 16:9 aspect ratio fits the player best. Upload in 4k when possible since video resolution is ever increasing. For Shorts, use a 9:16 format in 1080p

YouTube is by far the most popular video sharing website so far, and with Google's backing, will continue to be. With a basic account, you can upload videos smaller than 128GB and less than 15 minutes in length. You can upload as many videos as you like. Once your channel gets a following of 1000 subscribers and 4000 hours watched in the last 12 months, you can even earn some extra cash from your videos by monetizing your channel, especially if one of the videos goes viral. YouTube also has a password-protected private setting if you just want to share videos with your closest friends.

Facebook - *Social Media Video Sharing*
Facebook also allows you to upload 360 videos to your account. The maximum recommended video size is 10 GB with a max stream duration of less than 240 minutes. Facebook obviously also allows standard HD video upload as well for your Single Lens Mode videos. You can also post directly from the Insta360 App to Facebook.

Vimeo - *Creative/Non-commercial Videos*
Vimeo also supports 360 and standard video content. With a free basic account, you can upload 500MB of video per week up to 5GB total storage. The content must be original and non-commercial. You also have the option of making your videos private with password protection, which is convenient if you want to share selected videos with a limited crew.

Veer – *360 Video Sharing for Experienced Creators if Accepted*
Veer.tv used to offer hosting their online hosting services to anyone but has shifted over to hosting content from experienced creators only. As your knowledge grows and 360 video content improves, consider creating a portfolio and applying. They offer a variety of features and perks once you become an accepted creator.

REFRAMED VIDEO SHARING

Instagram – *Of course, one of the most popular. Host Single Lens Videos in multiple formats.*
Instagram does not yet host 360 videos, but there are lots of formats for sharing videos. Video is a super effective medium on Instagram and the X3 is one of the best tools for creating a wide variety of Instagram content. For IGTV, Instagram Stories, and

Instagram Live, reframe your videos to 9:16 vertical. For Feed videos, Instagram supports anywhere from 1.91:1 to 4:5. 1:1 square videos or 16:9 fall into that category for reframing in Insta360's apps.

TikTok – *The flashy X3 content is perfect for Tik Tok, where almost anything goes.*
Post your reframed 9:16 videos to Tik Tok. Trick shots (Shot Labs for example) and quick, snappy edits are best to grab attention. And since 1080p is all you need for resolution, everything can be edited using the Insta360 App.

360 PHOTO-SHARING

Facebook
Facebook allows easy 360 photo sharing. After uploading a 360 photo to Facebook, the sphere icon in the corner indicates that Facebook automatically recognized your file as spherical content. If your 360 files are not showing up in 360, the metadata probably just needs to be reinjected. Obviously, reframed photos can be shared on Facebook as well.

Instagram
Instagram doesn't yet support immersive 360 photos, but there are still a few creative ways you can show off your 360 content.

• First, you can edit your photo and save it as a flat image, such as a tiny planet, or a standard wide angle image pulled from a 360 photo.

• Second, edit a short 360 video animating one of your 360 photos using the Insta360 app. The user won't be able to navigate your image, but you can still share the magic of 360.

• Third, use Instagram's carousel album feature to share a swipeable horizontal panorama. There are several apps available to easily slice your panoramas into multiple photos to use with the carousel. On iOS, use Unsquared or Panorasplit. For Android, use PanoramaCrop. If you want a pano to stand out, upload the full panorama first, followed by the sliced images.

Google Photos
Google automatically recognizes 360 photos. Upload your photos to Google Photos, click on Share and get a link to share your photos.

After uploading your 360 photos to Google Photos, you can also choose to upload your 360 photos to Google Street View on Google Maps. After uploading fifty 360 photos (listed privately), you can get listed as Trusted Street View Photographer, which can be a fun travel hobby with a few perks.

Upload to a 360 Photo Sharing Site
You can also choose to upload your 360 photos to a photo sharing site, such as Kuula.co, Orbix360 or RoundMe. Some of these sites allow you to share the link or embed the photo on your site.

• Kuula.co offers a wide variety of 360 hosting services, including adding filters, simulated flare, 360 graphics and illustrations to your photos. Some plans charge a fee, but you can get started with the Free plan to start sharing photos.

• Orbix360 is another free 360 sharing site which allows you to create virtual reality experiences, including uploading 360 photos.

• RoundMe is a fee-based sharing platform for creating Virtual Tours around the world. You can see what other people have created and get started with their free account if you want to contribute.

VR HEADSET VIEWING

If you own a VR headset or want to create an experience for VR headset users, there are several ways to host your content to be compatible with a headset. 360 videos hosted on Facebook can be viewed using Oculus headsets, which include Oculus, Rift, Gear VR and Oculus Go.

VR headset viewing through YouTube is only compatible with Daydream, or PlayStation (VR). The playback quality through YouTube is higher than Facebook. You can also view YouTube videos in VR using Google Cardboard, which splits the playback screen into two smaller screens using your phone and a cardboard viewer. Google Cardboard is a cheap and easy way to experience VR viewing with your phone.

Veer.tv offers VR photo and video content for all headsets.

WEBSITE EMBEDDING

If you run your own website or blog, you can use a 360 hosting site to embed 360 photos and videos on your site.

The following sites are four free options for embedding 360 content on your site using an embed code: YouTube, Facebook, Orbix360 and Kuula. After uploading your videos or photos, use the Share feature to copy an embed code, which can be added to your site.

VIRTUAL TOURS

Virtual tours are one of the most useful ways to share your 360 experience with other people. We already covered some of the best tools for creating virtual tours in the Real Estate tips in Step 4.

PHOTO PRINTING

As you have learned, the Insta360 X3 takes high-resolution photos that allow for a wide variety of printing options. Printing flattened 360 imagery is the easiest way to showcase your newfound photographic skills, but there is also an interesting option available for printing spherical photos, which is shown at the end of this list.

With the plethora of creative photo printing options available, here are a few of the best for printing flat photos:

ACRYLIC FACE MOUNTS

The printed photograph is mounted behind a thin layer of acrylic and usually comes with a hanger mounted to the back, so it is ready to hang. Acrylic Face Mounting creates a modern look for wall art.

It's a tricky process, but Bumblejax, Shutterfly, and BayPhoto are a few labs that can produce these eye-popping prints for you.

METAL PRINTS

This printing process actually prints your photo directly onto aluminum, creating a unique modern display, with lots of shine and realism.

Google "Metal Prints" or check out BayPhoto, Shutterfly, or Bumblejax.

WOOD PRINTS

Photos printed on wood have a very organic feel and create an instant art piece. If you print without a white under layer, the wood grain shows through your image, giving it a unique texture that is perfect for adding a personal touch to your photography.

Check out Woodsnap.com, Shutterfly, or BayPhoto, or Google "Wood Photo Prints" to find a printer who can print your photos on wood.

GALLERY WRAPS/GICLEES

Make your photo look like a traditional wrapped piece of art by printing it on a gallery wrap. Choose Metallic Photo Paper for extra vibrancy in your photo.

Google "Gallery Wraps" or check out BayPhoto for printing options.

SPHERICAL PHOTO PRINTING

For a different kind of art piece, you can print your 360 photos onto a sphere. Snapspheres will do it for you, but their printed spheres are very expensive.

CONGRATULATIONS, YOU NOW HAVE THE KNOWLEDGE AND POWER TO CREATE AND SHARE YOUR EXPERIENCES WITH THE WORLD! IN STEP 7, YOU WILL LEARN MORE TRICKS TO TAKE YOUR INSTA360 FOOTAGE TO THE NEXT LEVEL.

STEP SEVEN

BEYOND THE BASICS

Take It Further

Once you have mastered the basics of using your X3 camera and see how much fun this camera can be, you will probably want to add more flair and style to your videos and photos. When you are ready to add some extra features to your camera setup, these additional tips and accessories will help you get even better shots.

The list of accessories and extras that are available for your Insta360 X3 goes on and on, but in this step, we will break it down to the accessories that will make the biggest impact on your footage and on your filming experience.

THESE ACCESSORIES WILL INSTANTLY IMPROVE THE QUALITY OF YOUR VIDEOS AND PHOTOS:

THE X3 UNDERWATER

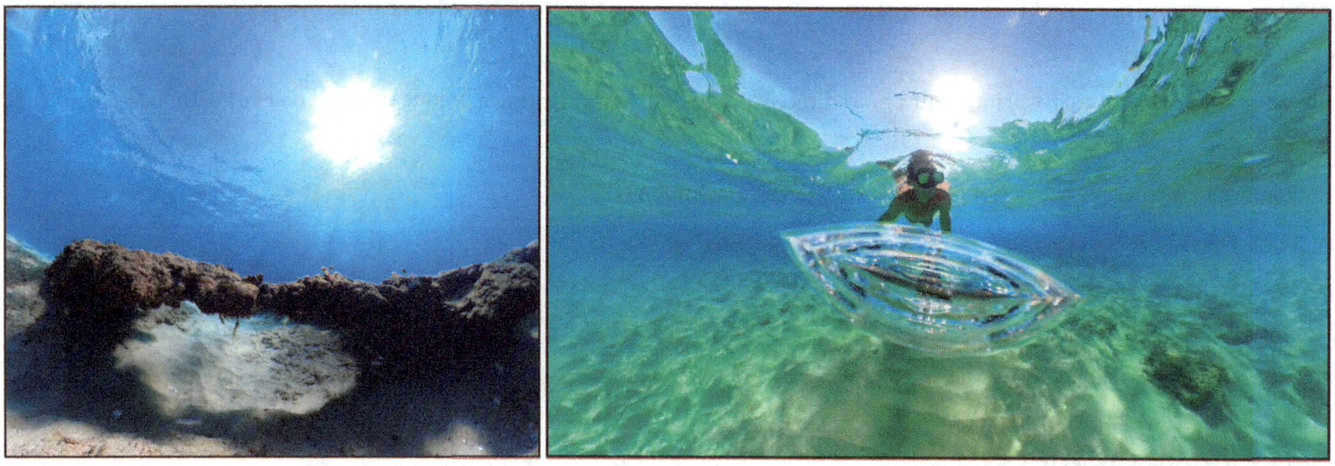

The image on the right shows the area of the frame that is blocked by the Dive Case. Take this into consideration while filming with the Dive Case underwater.

When most people think of 360 cameras, they probably don't think about underwater footage. Think again, because when the X3 is used the Dive Case, the results can be mind bending.

Even though the X3 is waterproof without a case, the curved lenses don't produce clear imagery underwater. However, when you put the X3 into the Dive Housing, you now have crystal clear shots underwater. This opens up possibilities for underwater VR experiences and unique perspectives.

Using the X3 with the Dive Case can be tricky at first because the base of the Dive Housing blocks out the shot. When using the Dive Case, **make sure your subject is to the side or at the top of the case.** Then, select the stitching for the Dive Case. Insta360 also offers **AquaVision 2.0**, which is automatic color grading for underwater footage. If you are a water lover, get under the water and see what you can create.

A quick tip though, if you are recording above the water, it's often easier to use the camera without the Dive Case.

PORTABLE LIGHTING

Most of your daytime lighting will come from the sun, but using an external light can extend your filming time or even just fill in light when the sunlight is not bright enough.

Mounting an external light directly to the Insta360 X3 is tricky, especially for filming in 360. The **size of the light needs to match the depth or be narrower than the X3 camera** so it doesn't show up in your videos or photos. Insta360 makes the Utility Frame, which holds the X3 camera and has a cold shoe mount on the side for mounting a light to the camera.

There are a variety of external lights that can be mounted directly to your X3 or placed in a scene for spot lighting. Due the compact size which matches the depth of the X3, **GoPro's Light Mod** is the best choice for an external portable light.

LumeCube also makes some useful lighting options for vlogging. Traditional LumeCubes are too bulky to mount onto the Insta360 X3, but the **LumeCube Panel Light and Panel Mini** are thin enough to mount on top of the X3. These lights offer white balance settings and provide more light than GoPro's Light Mod, but they are not waterproof.

When using lights when filming 360, **use the Isolated Exposure setting** to set the exposure for the lens where the light is pointing since this will be your main focal point.

FOR NIGHT VIDEO

Mount a light to your X3 for night filming. Night video filmed without artificial lighting requires using a high ISO setting resulting in lots of "noise". The use of a light allows you to bring your ISO back down a few stops, improve the view of the scene and continue filming into the night.

AS A FILL LIGHT

In low light settings, such as under a canopy of trees or on a cloudy day, external lights can be used to fill in shadowed areas. Place the light up against a subject or one of the key landscape areas for highlighting. Or place the light behind a translucent object for backlighting.

NOW THAT YOU KNOW WHICH ACCESSORIES WILL IMPROVE THE LOOK OF YOUR FOOTAGE, CHECK OUT THESE TOOLS TO ENHANCE YOUR FILMING EXPERIENCE.

GPS ACTION REMOTE CONTROL

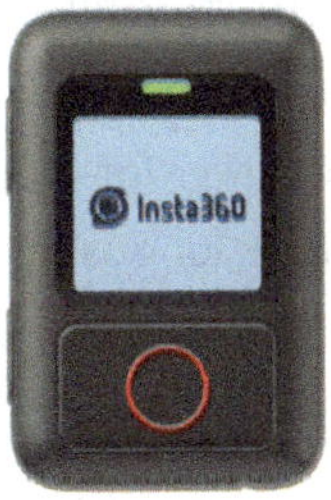

The GPS Action Remote can be used with the Insta360 X3 to control your camera remotely. Taking control beyond your phone's App, this waterproof remote really opens up the possibilities for different shots.

The Smart WiFi Remote is IXP8 waterproof and can control your camera from up to 65'/20m away in the right conditions away which makes it much easier to disappear from your 360 shots.

The Remote also **adds GPS data** to your X3 videos for sports and travel loves who want to show their stats overlaid onto their videos.

Try out these uses for a Remote:

SCENIC SHOTS AND REAL ESTATE PHOTOGRAPHY

Set up your camera and compose your shot. Then you can walk away and press the Shutter Button when you are out of the camera's view. The extra distance allows you to film a wider range of scenes without appearing in your shots.

ACTION PHOTOS

When shooting self-portrait action photos, the remote is especially useful. With video, you can record continuously during the action, but with self-portrait action photos, you need to push the Shutter Button at the exact right moment even when the camera is mounted at the end of a long pole.

INACCESSIBLE ANGLES

If your camera is mounted somewhere inaccessible, like on the wing of an airplane or on the side of your truck while you are off-roading to some remote camping spot, use the remote to start and stop the video during the exciting parts of the ride. Depending on the distance between you and your camera, you may also be to use the Insta360 App on your phone for this type of remote access.

UNDERCOVER ANGLES

If you are trying to record footage without anyone noticing, set your camera in the right spot and start recording or taking photos as soon as the action starts. This works great for nature shots because animals are more likely to approach your camera when no one is around.

OFF THE GRID X3 SETUP

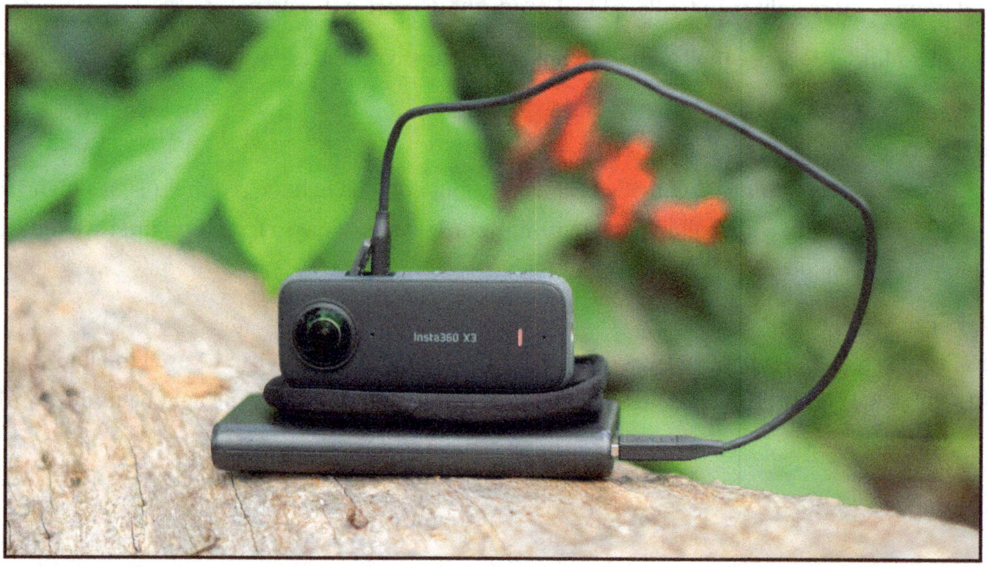

Action camera enthusiasts live outside of the box. With these accessories, you can step away from a wall socket and your computer for days or weeks and still be able to record your adventures:

PORTABLE BATTERY CHARGER

Out of the box, the only way to charge the X3 is to plug it into your computer or wall charger and wait for the batteries to refill. The Insta360 X3 Fast Charge Hub lets you recharge up to 3 batteries at a time with a USB port or wall socket giving you the freedom to get out there and disconnect. A portable power bank, like one by Ravpower or Iniu, is a great source of power for the dual battery charger. You can also connect any portable USB charger or USB phone charger directly to your camera as long as it outputs 5V and 3 Amps.

For real freedom from the grid for long trips through rural areas, a portable USB solar charger like the HiGoing 14000mAh Solar Power Bank (5V/2A) will be your ticket to ride.

PORTABLE DATA STORAGE

If you know you are going to be capturing lots of footage without being able to offload onto a computer, the best option is to bring a few high capacity (128 to 1TB) microSD cards. When you fill one up, swap it out for a new one and keep it in a safe spot. Another option is to use a portable memory card backup device to store your photos and videos if you want to travel light. The Gnarbox Portable Backup & Editing System pretty much eliminates the need to travel with a laptop, giving you storage and editing capabilities in one portable device for the content creator who really wants to roam free. This will give you plenty of storage to empty your memory card, get some editing done in your down time and keep filming as you wander on.

NOW GET OUT THERE AND HAVE FUN!

TROUBLESHOOTING

These are some of the most common problems Insta360 users have run into with previous camera models. If you encounter any of these problems, try these solutions first. You can always contact Insta360's customer support on their website, and they will troubleshoot any problems with you right away. See Insta360's website for their customer service contact info for your country.

CAMERA MALFUNCTIONS

If you experience any of the following problems: first, **try the solution offered.** If the problem happens repeatedly, **reinstall your camera's firmware** by performing a manual update through Insta360's website. It's possible that the firmware did not install properly during the initial installation. If the problem persists, contact Insta360's customer support.

Problem: The camera heats up when recording.
Solution: This is normal, especially when filming at high resolutions or frame rates. If you continue to experience overheating, temporarily turn off your camera to let it cool down. Also, recording in short stints or at a lower resolution will reduce the strain on your camera.

Problem: The camera freezes up and stops responding.
Solution: This is not normal and could indicate a problem with the communication with the microSD card. To unfreeze your camera, remove and reinsert the battery. Make sure to use a compatible microSD card with a large storage capacity to prevent using a maxed-out card. If it continues to happen, try manually reinstalling the firmware through the Support page on Insta360's website.

PLAYBACK ISSUES

Problem: Your computer does not recognize the memory card when your camera is connected to your computer.
Solution: Make sure your computer is running the current operating system.

For Mac Users: Some programs may interfere with the communication between your camera and computer. If the computer is not automatically recognizing your memory card, you can also import the footage using Image Capture, which is located in your Applications.

Problem: There is only audio and no picture.
Solution: First, check your video player or editing software. The .mp4 files are also used for audio files, so some programs may not interpret the codec correctly and recognize the file as an audio file. After verifying that your editor is reading the file correctly, reinstall your camera's firmware by performing an update through Insta360's website, through the Insta360 App. If the problem continues, contact Insta360's customer support. You may need to get a replacement camera.

If you are experiencing any other issues with your Insta360 X3 or any of Insta360's other products, contact their Support team.

Made in United States
North Haven, CT
11 November 2026

22095608R00070